T0133504

Security in IoT-Enabled Spaces

Security in IoT-Enabled Spaces

Fadi Al-Turjman

CRC Press
Taylor & Francis Group
Boca Raton London New York

CRC Press is an imprint of the
Taylor & Francis Group, an **informa** business

CRC Press
Taylor & Francis Group
6000 Broken Sound Parkway NW, Suite 300
Boca Raton, FL 33487-2742

© 2019 by Taylor & Francis Group, LLC
CRC Press is an imprint of Taylor & Francis Group, an Informa business

No claim to original U.S. Government works

Printed on acid-free paper

International Standard Book Number-13: 978-0-367-11123-6 (Hardback)

To my dearest parents, my brother, and my sisters.

To my wonderful wife and my little stars.

Contents

Preface

The Internet of Things (IoT) is a system of connected computing devices, objects, and/or people that are provided with unique identifiers to transfer data over a computer/cellular network wirelessly. And thus, security and smart spaces are considered one of the most significant topics in IoT nowadays. The implementation of secured smart spaces is at the heart of this concept, and its development is a key issue in next-generation IoT. This book is dedicated to addressing major security aspects and challenges in realizing smart spaces and sensing platforms in critical Cloud and IoT applications. Challenges vary from security attacks to privacy and reliability issues in safety-related networks. The aim of this book is hence to focus on both the design and implementation aspects of security models/strategies in smart spaces which are enabled by wireless sensor networks and Radio-Frequency IDentification (RFID) systems. It will mainly focus on seamless data access approaches and encryption/decryption aspects in reliable IoT systems.

By Fadi Al-Turjman

Author

Fadi Al-Turjman received his PhD degree in computer science from Queen's University, Canada, in 2011. He is a professor with Antalya Bilim University, Turkey. He is a leading authority in the areas of smart/cognitive, wireless and mobile networks' architectures, protocols, deployments, and performance evaluation. His record spans over 180 publications in journals, conferences, patents, books, and book chapters, in addition to numerous keynotes and plenary talks at flagship venues. He has authored/edited more than 12 published books about cognition, security, and wireless sensor networks' deployments in smart environments with CRC Press/Taylor & Francis Group, and Springer (Top tier publishers in the area). He was a recipient of several recognitions and best papers' awards at top international conferences. He led a number of international symposia and workshops in flagship IEEE ComSoc conferences. He is serving as the Lead Guest Editor in several journals, including *IET Wireless Sensor Systems* and *Sensors, Sensors* (MDPI), and *Internet of Things* (Elsevier).

1
INTRODUCTION

Contents

I. Introduction

Personal and social smart spaces are fields where Internet of Things (IoT) has been employed and has made some improvements in these areas. IoT has impacted fields such as social networking where the vast connection of smart devices has enabled users to connect and interact without worrying about the distance between them. Historical queries represent another field where anyone can retrieve or store information about anything from anywhere without distressing about losing that information. Moreover, IoT has made an impact in the security sector where the installation of smart security systems is used to guard property and prevent theft.

Since in general enabling technologies have restricted authentication privileges for mobile users, different strategies are introduced for the extension of user authentication over IoT-based environments. Commercialization of remote applications and security issues in femtocells have gained much attention of the researchers to satisfy the security properties of authentication and key agreement protocols. In general, the development of security protocols is more challenging and should also consider mitigation of the computation and communication cost. Moreover, considering the femtocell and macrocell energy consumption for heterogeneous networks is another key challenge. And covering a large area with the optimized deployment of macrocell and many overlapping femtocells needs a careful consideration in order to realize secure femtocells in the IoT era.

Lately, cloud computing (CC) and IoT-based techniques have been emerged with the wireless networks for the purpose of storage and data access at any time over the Internet. In the cloud, the user can find the set of hardware devices, network connections, storage spaces, data services and application interfaces that are easily accessed over the Internet. As the CC has become a prominent research topic among the researchers, user authentication schemes are highly recommended for the purpose of user authentication, authorization and accounting while the cloud services are in use. As a result, secure data access is necessitated between the source of information and the associated mobile/static nodes to assure that the collected data is securely exchanged with the legitimate user.

This book presents authentication and session key agreement methods using bilinear pairing and other interesting techniques. The objective is to provide mutual authentication, session-key agreement, data confidentiality, and resilience to node-capture attacks in smart spaces. We overview the advantages for IoT-based algorithms and approach in smart environments towards further improving the quality of our daily life while spotting the light on key design challenges in cybersecurity. Accordingly, our main contributions in this work can be summarized as follows:

- We start by overviewing the importance of smart spaces nowadays and their enabling technologies. We highlight open research issues and possible strategies that can be applied for further optimization in performance.
- Various techniques and tools available for the mobile smart spaces applications, such as the eHealth, safety, and smart vehicular networks in IoT environments, are also presented in order to realize more secure solutions. These solutions and security challenges are discussed in detail.
- We describe prominent performance metrics in order to understand how the data access is secured and personal privacy is protected in the IoT era.

II. Book Outline

The rest of this book is organized as follows. In Chapter 2, we delve into an overview for open research issues in smart spaces. In Chapter 3, we overview prominent security issues and threats in smart vehicular networks and wireless communication protocols. Chapters 4 and 5 provide novel security approaches and frameworks for the wireless sensor networks' applications in smart spaces. Chapter 6 proposes a security model for IoT virtual networks while highlighting technical issues and design factors in their communication protocols. Chapter 7 emphasizes the reliability issue in secured edge spaces and suggests a new IoT-based Edge Framework. In Chapter 8, we propose a context-sensitive approach for data access in medical smart-spaces' mobile applications while satisfying guaranteed security levels. In Chapters 9 and 10, we focus on secure data caching and key agreement approaches in the smart space.

2
SMART-SPACES IN IoT[1]

Contents

[1] Previously published in F. Al-Turjman, "5G-enabled Devices and Smart-Spaces in Social-IoT: An Overview", *Elsevier Future Generation Computer Systems*, 2017. DOI: 10.1016/j.future.2017.11.035.

Abstract

The abundance of smartphones, with their growing capabilities potentiates applications in numerous domains. A typical smartphone nowadays is equipped with an array of embedded sensors (e.g., GPS, accelerometers, gyroscopes, RFID readers, cameras, and microphones) along with different communication interfaces (e.g. Cellular, WiFi, Bluetooth, etc). Thus, a smartphone is a significant provider for sensory data that awaits the utilization in many critical applications. Primers of this vision have demonstrated success, both in the literature and application's market. In this literature review, we present the main motivations in carrying these smart devices, and the correlation between the user surrounding context and the application usage. We focus on context-awareness in smart systems and space discovery paradigms; online versus offline, the femtocell usage and energy aspects to be considered, and about the ongoing social IoT applications. Moreover, we highlight the most up-to-date open research issues in this area.

Keywords

Internet of Things (IoT), Smart environments, Sensors; 5G; Smartphones.

I. INTRODUCTION

In this information age, online activities form a significant part of our daily life. Social networks and collaborative sites such as Twitter, Facebook and Google+ are incredibly popular online forums, offering easy and compelling ways for millions of users to post content and interact with each other. In addition to providing attractive mediums for person-person interactions, social networks also offer unprecedented opportunities for social data analysis, i.e., big-picture views of what people are saying, because they contain a deluge of opinions, viewpoints, and conversations by millions of users, at a scale that would be impossible using traditional networks such as the internet and circuit switched telephone networks. The 5th Generation of wireless technologies, abbreviated as 5G, are the proposed next generation telecommunications standards for this kind of applications. 5G comes as the demand for better, and faster wireless connection. As stated in [1], by the time 5G is fully embraced, there will be tens or hundreds of billions of devices that require the use of 5G technology, not only because of personal usage, but also due to many new applications. One of the major new application that will contribute to this high number of devices is the highly anticipated Internet of Things (IoT). The term IoT has been loosely used in many scientific research areas as well as marketing and sales. According to [2], IoT can simply be defined as a dynamic global network infrastructure with self-configuring capabilities based on standard and interoperable communication protocols. There are estimates that 50 billion devices will be wirelessly connected to the Internet by 2020 [3]. Meanwhile, the slow but steadfast introduction of IPv6, in addition to the great proliferation of sensing and tagging technologies are speeding up the realization of the Internet of Things (IoT) – an Internet where everything is reachable and can communicate [4][5]. This means that a larger number of the familiar Wireless Networks will be deployed, generating great amounts of data to monitoring devices or personnel. This Ultra Large Scale communication paradigm will inevitably rise to become a dominant necessity for the various sections – both public and private. However, there are serious challenges in realizing such vision. Traditional design and deployment of wireless networks has been application-specific, that is, single-purpose networks [6]. The tradition stemmed from the increasing cost boundaries and the diminishing margins of practical feasibility when deploying practical, real-life wireless networks deployments, especially maintenance of cellular functionalities in very dense deployments or spread over large geographic areas where the small cells (femtocells) in 5G plays a key role [5][7].

As the computation and the communication process of heterogeneity hidden networks involve intelligent decision making, the human-to-machine perception on the realization of large-scale IoT environment is bridged as a key challenge for Social Internet of Things (SIoT) [8][9] in literature. In this chapter, SIoT is represented as a system of smart things, which allow people to interact with each other to share / exchange social information. Though the SIoT can improvise the interaction and the navigation between the communicated objects, the contextual data of the objects should be handled with respect to situation awareness. The SIoT classifies the contextual data as objective and subjective [10]. The former context is used to define the physical aspects of sensing objects, such as device status, time, location, service availability etc. The latter is used to represent the short-term goals, trustworthiness and preferences. So far, the combination of both has not been studied for SIoT in terms of security [11]. And while this view continues to thrive in the literature, the arguments for the boundaries and margins are no longer valid [12]. As well, disappointments resulting from several practical deployments based on this design view have raised concerns for the viability as a sustained design approach [13][14]. Ultra large scale networks, such as the cellular networks which depends significantly nowadays on the smartphones, shall respond to increasing activity in a recent direction of research advocating public sensing. In public sensing, systems probe resources like cell phones, laptops and web servers to build a virtual network on the Internet – one which could be remotely queried for information collection [15]. These systems were proposed for promoting activity monitoring [16], air quality monitoring [8], social networking and, using neural signals, controlling mobile phones [9]. These approaches can be either participatory or opportunistic, offering great flexibility in both design and adoption. This depends heavily on the smartphone utilized sensors.

Accordingly, the main contributions in this survey article can be summarized as follows. A comprehensive background about the 5G standards and their IoT-specific applications are outlined while emphasizing energy consumption and context-awareness. The recent directions in using smartphones' sensors with alternative design approaches that can functionally contribute to more scalable operations in smart social spaces are overviewed. Online versus offline mobility detection applications, and potential communication technology, namely the femtocell, which can be a strong candidate for the IoT paradigm realization in practice have been investigated. Modeling and energy aware metrics are outlined as well. Moreover, key open research issues are highlighted and discussed. In order to assist the readers, we provide in Table 2.1 a list of acronyms along with brief definitions as used throughout this article.

Table 2.1 Acronyms and Definitions

ACRONYM	DEFINITION
IoT	Internet of Things
UE	User Equipment
RFID	Radio Frequency Identification
HAN	Home Area Network
NAN	Neighborhood Area Network
ROF	Radio-Over-Fiber
ICT	Information and Communication Technologies
RAN	Radio Access Network
QoS	Quality of Service
MBS	Macrocellular Base Station
RNC	Radio Network Controller
FAP	Femtocell Access Point
LTE	Long Term Evolution
CVCs	Context-aware Vehicular Cyberphysical systems
WBAN	Wireless Body Area Networks
MCC	Mobile Cloud Computing
EE	Energy-Efficient
BS	Base Station
PA	Power Amplifier
SE	Spectral Efficiency
HCSNet	Heterogeneous Cloud Small-cell Network
UDCSNet	Ultra Dense Cloud Small-cell Network

II. SMARTPHONE USAGE AND CONTEXT-AWARENESS

Mobile phone usage statistics can be obtained in different ways. One approach is asking user to manually log their activities and surveys [17]. Second approach is collecting data from the device with an agent application [18]. Third is combining the first and second approach [19]. However, user reported data can be prone to errors, due to personal biases and limitations, etc. Therefore a better approach is to apply context or semantic analyzing methods over collected mobile sensor data [20][21]. Context-aware systems need to collect a variety of information about the user's current status and activities, some of which may be regarded as personal and make it user-centric system [22][23]. We have made a few attempts towards realizing context-aware networks [24]. We implemented a Smart Spaces platform, called "CAR" that integrates smart spaces with social networks through the IP multimedia subsystem; creating truly context-aware and adaptive spaces. We designed and implemented all components of the CAR including the central server, the location management system, social network interfacing components, service delivery server and user agents. Li and others presented an analysis of app usage behaviors by using a famous Android app marketplace in China, called Wandoujia [25]. They studied over 0.2 million Android apps and 0.8 million users. Their findings on usage patterns include app popularity, app category, app selection and network usage [25]. However, this work is done on a limited geographical area and does not give a global perspective.

It is elemental in the design of context-aware solutions to ensure attractiveness by ensuring interoperability to existing standards. Consider, for example, the case of user profiles in social network services such as facebook. The authorized extraction and management of such profiles is elemental to any context-aware framework. At certain stages of a context-aware framework, it will be unavoidable to resort to capable profile extractions such as FOAF (Friend of a Friend) as it might prove useful in this specific domain. Briefly explained, FOAF is a decentralized semantic web technology, and has been designed to allow for integration of data across a variety of applications, websites, services, and software systems. However, careful investigations are required in order to verify its appropriateness to fulfill the above noted generic characterization requirements.

Application as a Service (AppaaS) described in [26] provides an overview of the most appropriate system architecture we can think of for smart spaces utilizing smartphones. The architecture involves the user device loaded with an AppaaS mobile application, in addition to a space/context management server. The AppaaS mobile application comprises of a graphical user interface (GUI) which is used to take inputs from users, a Space Handler which collects different context information from users and Service Delivery makes sure that services are delivered in the form of applications relevant to the user's context. AppaaS mobile application provides users with an interface which can be used to provide certain inputs. When a first time user registers with AppaaS, the user provides basic information and the information is sent to the AppaaS server. Upon successful registration, the user is taken to a login screen, here the user provides the login information and upon successful login, user's relevant scheduling information is fetched from AppaaS server. Obviously, the above described context-aware setup depends greatly on facilitating state preservation, which essentially refers to applications able to save their latest status and user-specific data for future access. Current mobile platforms do not natively support state preservation at any level. However individual applications can manage their own state at different levels depending on the objective from state preservation. For example, applications may use checkpointing techniques [20][27][28] to suspend and resume their execution or for migration purposes. Generally speaking, an application-independent state preservation of user-specific data remains challenging. To this end, new paradigms shall assume that applications would provide two proprietary APIs for the sake of state preservation. One API saves the application current state, where all user-specific data is saved in an XML file format. The other API uploads an application state from an XML file when the application launches.

In general, wearable sensors are promising technologies in the field of user behavior analysis and monitoring. In the near future, several mobile applications will be dependent on these tiny wearable devices, through which user habits, user mobility, application usage and context awareness can be further investigated. In Table 2.2, we tabulate the reviewed attempts in this regards.

Table 2.2   Comparison of the Context-Aware Usage in Mobile Applications

REFERENCE	CONTEXT RESOURCE	METHOD
[25]	Statistics of Apps w.r.t. management, device price, foreground vs background app with network connection.	8M users, 260K devices.
[27]	Location (via WiFi), app duration and type.	14 teenagers, 4 months, HTC Wizard, logging and interview.
[20]	Stay points (GPS, WiFi) and labelling, app type and voice/sms, density.	77 participants in 9 months, using Nokia N95.
[18]	location, interaction frequency, battery use, connectivity, call frequency.	1277 types of devices, 16K participants, 2 years 175 countries, using Android.
[29]	Number of user interaction, network traffic, app use, and energy drain.	255 (33 android+ 222 windows phone), using an app for logging.
[28]	Battery, location, time, selected apps using app sequence histories, and connectivity.	50 participants in 9 months.
[30]	Stay points (GPS), app duration, <loc,app>sequences.	30 students in 3 months.
[19]	App frequency and type.	28 participants in 6 weeks, using an app for statistics.
[31]	Location: GPS, Bluetooth, WiFi, accelerometers.	8 participants in 5 months, using an agent on device.

III. SMART SPACE DISCOVERY

In this section we elaborate on the most significant components in the smartphone usage, which is the space discovery paradigms. Moreover, we categorize the existing tracking Apps in this area.

A. Space Discovery Paradigms

Facilitating location based services is fundamental to the realization of the efficient smartphone usage. For outdoors and open areas with direct line of sight, it is possible to depend on GPS or even terrestrial localization schemes, e.g., as offered by cellular networks. Announcing the location for indoor devices, however, becomes more problematic. In this direction researchers investigate the viability and efficiency of a multitude of space discovery mechanisms, to facilitate rapidly identifiable location-based services. Prominent contenders in this investigation include RSSI models, Near Field Communication (NFC), and Wi-Fi/cellular based models. Other investigations encompass their efficiency in indoor environments, operability under varying density constraints, privacy preservation, and distance to identification devices. For example, RFID readers for NFCs.

A prominent and novel direction of investigation encompasses employing Ultrasound based detection mechanisms. The distinct advantages of ultrasound communication include controlled transmission range, limited locale (no wall penetration), does not require additional hardware on part of the user, and has modest implementation requirements. More importantly, the behavior of ultrasound is more predictable than IR and other radio waves. Prominent disadvantages to ultrasound communication such as air speed and scattering could be addressed by careful signaling design, e.g. longer pulses. Careful considerations also need to be made when utilizing ultrasound communication in medical environments; all of which are core parameters to this research direction. To further elaborate, the intended operation of the proposed system entails sending information through beacon devices in the non-audible, ultrasound range (i.e., > 20kHz). This information can be picked up and interpreted by the user's device microphone. This setup readily indicates the wide feasibility of implementation given that any smartphone device would have a microphone. Furthermore, a few evaluations indicate the readiness of common smartphones to this. For example, Figure 2.1 shows the response of the Samsung Galaxy II for a continuous tone at 20 kHz [32]. It should be noted, however, that as the audio profiles of the device microphones vary across devices, a calibrator module might be required in order to prepare individual smartphones for this communication.

As a prototype, the beacon device can be implemented on a single-board computer can be used inside a beacon. The computer shall have enough computing power to playback an audio file with a sampling rate high enough to produce frequencies higher than 20 kHz and a wireless channel, such as WiFi, to allow for remote configuration. (Note that to be able to produce frequencies up to 21 kHz we need to generate the audio signals with a 44.1 kHz of sampling rate.) Such a beacon can encode the service code on its own board without the need of external CPU processing. Meanwhile, beacon devices will be designed to support more than one "audio tagged" device simultaneously. To achieve

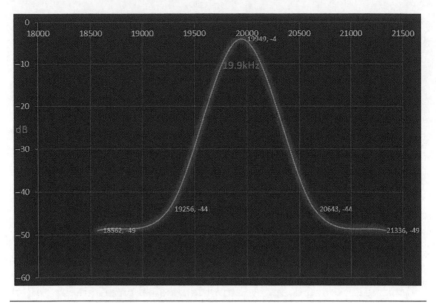

Figure 2.1   Frequency response of Samsung for the 20 kHz [32].

this, synchronization is needed between beacon devices in the same vicinity. On the client side, it is worth noting that a minimum sampling rate of 44.1 kHz is supported by most microphones, although a few experimentations indicate that ultrasound communication would work between if a sampling rate of 48 kHz was utilized. Moreover, the ultrasound system will be amended by a user configuration interface that facilitates visualization of designed locales, and aids the designer in placing the beacon devices. Figure 2.2 shows an instance of the configuration module communicating with the beacon devices for initial configuration download.

B. Online Tracking Apps

There have been a few attempts on getting user location for different purposes. Some of them include predicting application usage [28][30], friend recommendation systems [33], classifying applications or user characteristics [24]. Even though movement and location analysis are different, they share some common characteristics. First we need to answer which applications are used when users are mobile. We need to identify user movement for this purpose. We can classify the work

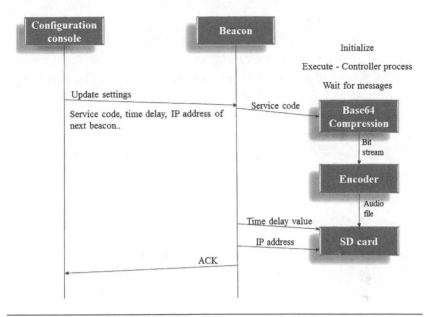

Figure 2.2 Initial configuration download.

done in this area to online and offline tracking applications. Online applications need a stable internet connection to send collected data to a cloud based storage [34].

Wagner et al. used mobile phone providers' Call Data Records (CDRs) and Wi-Fi based location information in order to analyze the user's data [32]. They avoided GPS in order to reduce the battery consumption. Their application was installed on more than 16000 contributors in 175 countries. In this work it was found that most of the users turn off Wi-Fi connection to save battery. Also the majority of devices in their dataset saw 3 or fewer Wi-Fi access points most of time. In this work there is not any information about app usage [32].

C. Offline Tracking Apps

In this section we look at offline tracking applications. Rahmati and Zhong have conducted an analysis in which they installed logging software on mobile phones and distributed them to users for research [27]. They collected app usage and location relation and also movement information. However, this work lacks detailed analysis. Another approach is storing user and context data and then uploading them daily to a server [20]. Offline apps collect user related information and send them to the cloud or a central serve when user becomes online. Lu et al. proposed a method for predicting app usage using stay points and location information. They used semantic locations approach and created a database containing application launches and stay locations offline in the devices. Their prediction algorithm is designed to be online to predict the subsequent app needs of a user according to the previous locations and app launches [25].

D. Social Space Provisioning

A social space takes into account usually the user's identity and personal information, location and time information, relevant to the situation or activity the user is currently involved in. As well, relevant to user's space information, social spaces aim at providing the user with the most relevant services in the form of a smartphone application by customizing the behavior of the smartphone application(s) depending on the user's profile, as described above in the Application as a Service (AppaaS) cotext-aware system.

Several attempts in the literature have been applied to eliminate the user involvement in smart-spaces communications. System components described above adhere to this motivation, achieving a high degree of service portability as the user moves from one space to another. The ultimate motivation of the social IoT, however, is tailor service delivery to the characteristics of the wireless and mobile environment. These include intermittent connectivity, limited features (display, processing power, etc), limited battery, in addition to heterogeneity of required resources. And while dependence on device participation in moving between spaces adds considerable flexibility, it ultimately restrains the device's lifetime and, in turn, the user's mobile experience.

To truly instill global seamlessness in social service delivery, it becomes inevitable to engage the cloud in IoT paradigms. This spans both virtualized services and resources, in addition to in-network processing. More specifically, the range of services that involve mobile devices providing data are on the rise, spanning entertainment services, such as online social gaming and networking, to crowdsourcing, such as collaborative participatory sensing as well as services that can be offered on the fly, such as video streaming of a current event. However, the rich functionalities that such applications over increasingly demand resources beyond the capabilities of inherently resource-constrained devices. Such lack of resource matching places limitations on the type of functionality and services that can be offered, restraining users from taking full advantage of their mobility passion. Cloud computing, therefore, offers the possibility to unleash the full potential of mobile devices to provide reliable data services. This expands on the notion of the aforementioned Application as a Service (AppaaS) to become Space as a Service (SpaaS). As a basis, SpaaS unfolds the full potential of service integration between the device and the space, going beyond the integrated sensing environment discussed above. SpaaS also goes beyond user migration to cater to user mobility. In such instances, the user's space expands and reduces as he moves from, say, the home to the car to the office to the mall, etc. More importantly, however, is that Spaas extends profile-based space customization to full space mobility, explicitly enabling a form of space virtualization by which a user can bring a distant space in his or her direct presence.

The aforementioned cloud-assisted mobile service architecture involves four key entities: a user, a mobile device, a cloud, and a data provider, as shown in Figure 2.3. The user represents the service consumer. The mobile device represents a mobile service provider and

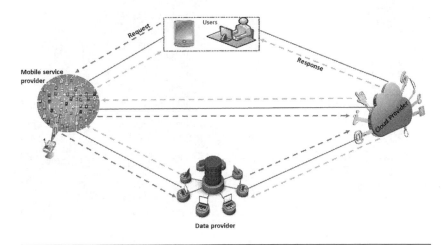

Figure 2.3 Architecture of SpaaS [26].

acts as the integration point where service execution plans are generated and decisions regarding offloading are made. The cloud is the supporting computing infrastructure that the mobile provider uses to offload resource-intensive tasks. Web service operations might involve third-party data processing during the execution of the service functionality, such as weather information or navigation databases. In such cases, data could be fetched from a data storage provider. In this architecture, the user sends the service request to the mobile provider. The mobile provider decides on the best execution plan and whether offloading is beneficial. The cloud offers elastic resource provisioning on demand to mobile providers. The mobile provider may collect the execution results from the cloud and generate a proper response for the user. It is also possible that the provider may delegate the cloud to forward the response directly to the user, given that no further processing is required at the mobile side.

The proposed framework encompasses the following major components, as depicted in Figure 2.4: Request/Response Handler, Context/Space Manager, Profiler, Execution Planner, Service Execution Engine, and Offloading Decision Module.

Briefly described, the profiler characterizes the offered space services operations and generates resource consumption profiles, while the

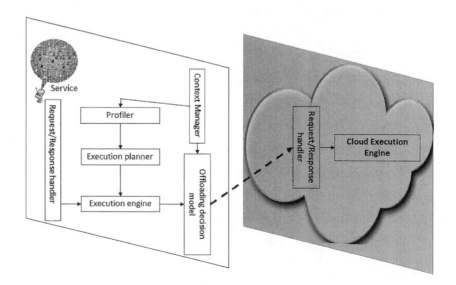

Figure 2.4 Details of SpaaS Components [26].

execution planner investigates all possible execution plans based on locations of required data and current context information. The service execution engine evaluates these plans and selects the best resource efficient plan that, in addition to satisfying the resource constraints, yields better performance and lower latency.

IV. SMARTPHONES & MOBILE APPLICATIONS TESTING

Mobile testing underwrites a flawless customer experience. No matter how amazing the smartphone mobile application is, if it has just one error discovered by the end-users, they will not be happy with it. Even worse, they will associate with the producer company or brand all similar kind of errors. That's why, companies pay lots of attention to test all the mobile application features and behaviors before it is on air. But testing all the features and behaviors of any application is impossible or unnecessary. Instead of this, automated testing strategies for mobile applications or test cases which are prepared according to these strategies can be used to save both time and cost. Also, mobile applications are different than desktop applications. In this section, we investigate these differences.

A. Mobile vs. Desktop and Web Application Testing

There are some fundamental differences and challenges between Mobile and Desktop Web application testing. If we understand these differences and challenges of the mobile testing, it will be easier to tackle them. Let us start with the definitions of application types. There are mainly three types of applications: Desktop, Web and Mobile. Firstly, Desktop application is a native application which executes on the user's local machine. This type of applications may have a network module to communicate with an external server. Secondly, the Web application is a kind of application that runs within a browser. All the information stored in an external server and browsers are used to communicate with it. Lastly, Mobile application called native application downloaded from an app store is intended to run on mobile devices such as the smartphones which are equipped mostly with iOS or Android operating systems. Mobile applications like desktop applications may have a network module to communicate with the external server.

Environments and Testing concerns of these application types are unique by nature. Mobile applications are inherently tied to the hardware and operating system. Therefore, they have more environmental situations and concerns than other application types. First, there are variety of mobile devices. According to the Google Play Store, there are 12402 types of mobile devices which runs on Android operating system. These devices differ in screen sizes and hardware capabilities. Also, there are variety of operating systems versions. There are 12 main versions of Android OS and 10 main versions of Apple iOS. Accordingly, almost every year a new version of these operating systems is introduced. Furthermore, there are variety of mobile network operators. There are over 400 mobile network operators in the world [35]. Each mobile operator supports different network technologies including LTE, CDMA, GSM, and some other local networking standards. In addition, mobile applications have some unique test cases, such as interruptions, battery consumption and global positioning system (GPS). However, while testing a mobile application, testers need to simulate an environment which considers these cases, this is the hardest part of mobile application testing. Accordingly, there are two ways to test these cases. The first way is Emulator testing. Testers need an Android or iOS emulator to do that, but simulating these features for all possible cases systematically is not easy because of the differences between real device and emulator. On the other hand, testers can use real devices to test the mobile application. This option provides more reality, but you cannot create an environment specifically for any situation.

B. Challenges of Mobile Applications Testing

Nowadays, there are nearly 13,000 types of mobile devices running 100 operating systems and versions serviced by 400 carriers worldwide [35]. If the goal is to test every permutation possible, then every test case will be run 9,100,000,000 times.

1. Device Fragmentation The most complicated aspect of mobile application testing is device fragmentation. The fragmentation is coming from mostly screen sizes and hardware capabilities. As a fact iOS device fragmentation is not a huge problem, but device fragmentation is an issue for the Android operating system. In 2012, there were 4,000 separate Android devices available. In 2016, the number is exploded to 12,000. Accordingly, it is not observed yet the total number of devices an app needs to be tested before releasing it in the market.

2. Operating Systems and Fragmentation Not just newer or smarter mobile devices, but also new versions of operating systems are being launched almost every year. The problem is compatibility issues that mobile applications face while being deployed across devices having different operating systems like Android, iOS, Windows, BlackBerry etc. or versions of an operating systems such as iOS 9.x, iOS 10.x.

3. Simulation Environment Mobile emulators and simulators are important and main method as a testing tool to verify the general functionality and perform regression testing. The testing is conducted in a simulated environment which is not real. Testers can set their environmental conditions while debugging the application manually or automatically.

A mobile application while functioning may face several interruptions like incoming text messages, incoming calls, incoming notifications, network coverage outage and recovery, battery removal or power cable insertion and removal. A well tested and developed application should be able to handle these interruptions by going into a suspended state and resuming afterwards. To do this, while setting simulation environment, testers needs to add these interruptions into the test cases.

The innovation in the battery usage duration field has not been quick as in the application consumption. End-user running and using lots of application during the day and some process are running on background without even noticing. These operations all requires a power and batteries tend to die. While testing applications, testers need to set not only possible battery percentages, but also battery usage. If application consumes too much power, this may cause less or not usage of the application.

V. 5G/FEMTOCELLS IN IoT SMART-SPACES

Given that IoT is strongly connected to the 5G future networking paradigm, in this chapter, we are spotting the light on the 5G network where the femtocell is a main player. Femtocell is a cellular network base station that connect standard mobile devices to a mobile operator's network using residential DSL, cable broadband connections, optical fiber or wireless last-mile technologies" [36]. It is an inexpensive compact base station providing equal radio access interface as a common macrocellular base station (MBS) towards User Equipment (UEs) [37]. It is a solution to offload from overloaded macrocells and increase the coverage area [36]. They are generally used for increasing indoor coverage and designed for use in a home or small business where there is a lack of cellular network or increasing the QoS (quality of service). It has advantages for both cellular operator and the smartphone user. For cellular operator, the main advantages stem from the increased coverage and capacity. Coverage area is widening due to eliminated loss of signals through buildings and capacity is increased by a reduction in total number of UEs that uses the macrocellular network. They use Internet instead of using cellular operator network. For customers, they have better service and improved coverage and signal strength since they are closer to the base station. Moreover, using femtocells leads to prolonged UE battery lifetime due to the close distance to the femtocell [36].

Typical femtocell structure consists of five main parts: femtocell device, DSL router, ISP (Internet link), mobile operator network and cellular tower (macrocell) [38] (as shown in Figure 2.5). Femtocell does not require a cellular core network since it contains RNC (Radio Network Controller) and all other network elements. It acts like a Wi-Fi access

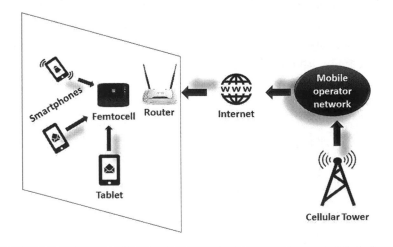

Figure 2.5 Typical Femtocell Structure [36].

point and it needs data connection to the DSL or Internet connected to a cellular operator core network [36]. Although it does not need to be under the macrocell coverage area, there are lots of examples for deployments in under the macrocell in order to increase the capacity and QoS in presence of huge user demands or significant shortage in coverage. Apart from this, it can be used in rural areas in order to provide cellular coverage where there is no macro coverage.

Although femtocell technology first designed to use indoor, there are lots of outdoor applications of femtocell technology. To illustrate, it can be deployed in transit systems such as bus, train, etc. In this application, mobile users connect to femtocell instead of macrocells or satellites. There is a transceiver connected to femtocell base station through wired connection and to macrocell or satellite through a wireless link [39]. Moreover, femtocells can be good solution to increase coverage and capacity in public outdoor areas, especially in crowded areas. The key point behind the femtocell is to bring cellular network closer to user and with this approach it manages to be low-power and low-cost technology [38]. It is usually difficult for a macrocell to provide indoor service since there is a signal loss. Moreover, 50% of voice calls and 70% of data calls comes from indoor [38]. There is an estimation that 10% of active femtocell household deployment can offload 50% of the overall macrocell load [40]. Thus, it increases revenues of cellular operators, and thus it is expected that there will be about 28 million unit of femtocell by 2019 [41].

Internet of Things (IoT) touches every facet in our life and our smart applications that can improve quality of our lives in different regions and domains such as home, highways, hospitals, etc. [42]. IoT describes communication capabilities of these objects with each other and to elaborate information perceived. IoT includes very wide range of applications (see Figure 2.6) and these can be grouped as: 1) Transportation and logistics, 2) Healthcare, 3) Smart environments, 4) Personal and social, and 5) Futuristic applications.

Femtocell can be used as a communication mechanism for the IoT paradigm, especially in the smart grid [43]. In [44] and [45], authors proposed the use of femtocells in a home area network (HAN) as a cost-effective technology. Health-care IoT applications are another example for using a femtocell in smart spaces. In [45], authors proposed an IoT-oriented healthcare monitoring system where sensors

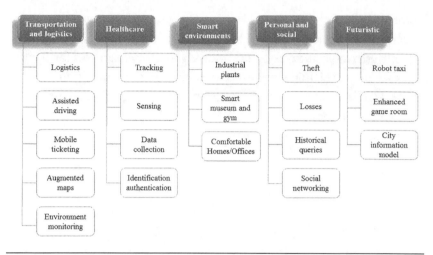

Figure 2.6 Application Domain of IoT.

gather together the particle measurements of an Android application and then they used LTE based femtocell network in order to send the data with new scheduling techniques. In [46], authors underline that femtocell is susceptible to Man-in-the-middle attacks when it is used to fix shadow area problem. Moreover, they propose the interlock protocol to protect the confidential information. In [47], the benefits of using 5G femtocells for supporting indoor generated IoT traffic is highlighted. Authors underline the fact that supporting the traffic produced from IoT is a major challenge for 5G and significant percentage of the traffic will be generated indoor.

Thus, it can be said that femtocells can be used in most of the IoT applications since there is a desperate need for a capable communication mechanism/tool. Indeed, it is expected that small cells such as the femtocell will be central to 5G network architectures both for human users and for IoT embedded systems.

VI. ENERGY ASPECTS

Energy is a key design factor in all enabling technologies nowadays which realizes the smart-space vision in IoT. Measuring energy in the targeted enabling technologies in this study including the smartphones and femtocells is of utmost importance towards realizing the smart spaces in practice. There are different energy-efficient (EE) metrics in the literature and they are applied at three different levels: the UE/component level, the base station (BS) level, and the network level. In component level, millions of floating point operations per second per watt (MFLOPS/W) and millions of instructions per second per watt (MIPS/W) can be used to calculate processor related energy consumption and the ratio of power amplifier (PA) output power to input power (ROI) can be used to calculate energy efficiency of power amplifier component. In base station level, EE metrics can be evaluated under two main categories. Bits per second per hertz per watt is for trade-off between energy consumption and spectral efficiency (SE) and bits times meters per second per hertz per watt is an energy efficiency metric when energy consumption, spectral efficiency and the transmission range of the base station is taken into consideration. In network level, obtained service relative to the consumed energy is evaluated by energy-efficient metrics which is power per area unit (watts per square meter) in order to evaluate the coverage energy efficiency. Table 2.3 gives the summary of the reviewed energy efficient metrics.

Since the structure of femtocells is similar to that of the macrocell, the macrocell component level energy metrics are also suitable for femtocells. However, for the BS and network level, the difference between the femtocell-supported and macrocell provided services and the interference between femtocells and macrocell should also be considered as femtocells are overlapped by a macrocell and use the same spectrum as the macrocell. In [49], authors proposed an energy efficient metric for the femtocell and macrocell heterogeneous

Table 2.3 Energy Efficiency (EE) Metrics

ENERGY EFFICIENCY METRICS	LEVELS	DESCRIPTIONS
ROI	Component level	Used to evaluate energy efficiency of power amplifiers
MIPS/W or MFLOPS/W	Component level	Used to calculate processing associated energy consumption
bits/s/Hz/W	BS level	For trade-off between energy consumption and spectral efficiency (SE)
(b*m)/s/Hz/W	BS level	Taking into consideration energy consumption, SE, and the transmission range of BSs
W/m^2	Network level	Used to evaluate the coverage EE

network that considers the service rate and power consumption in both femto-base stations and macro-base stations.

A. Energy Modelling

Generally, energy consumption of a wireless network [48] is evaluated at two different levels as described in Table 2.4. The first one, embodied energy consumption, calculates the total primary energy that is consumed for making the product. Embodied energy of for a femtocell for example is assumed as 162 MJ, same as a mobile terminal and the average life time of a femtocell can be assumed as 5 years. In other words, embodied energy of the femtocell per second is calculated as 1W [50]. The second level is operational energy consumption which can be defined as the amount of energy spent during a system's lifetime and it changes depending on different configurations such as age of the facility and load of the femtocell. In [50], average operational power of a femtocell is assumed as 6W. The total power consumption of a femtocell mainly depends on radio frequency power amplifier and the power amplifier of power supply.

In the literature, there are few models about the power consumption of a femtocell. On-off model is the most basic one which can be used for theoretical analysis where femtocell base station is assumed to consume unit power in active mode and zero power when it is off. However, it does not reflect actual power consumption. In [51], they proposed a linear power model that considers the traffic load. This model is used for analysis and more accurate than on-off model since the traffic load is defined. In [52], authors proposed a more detailed model and they argued that the traffic load has no effect on power consumption so it can be omitted.

In [53], authors proposed a simple analytic model to predict the femtocell base station power consumption based on offered load and datagram size. They tried to fit their model to real experiment that

Table 2.4 Energy Consumption of a Femtocell BS in Different Levels

ENERGY CONSUMPTION LEVEL	DEFINITION	AVERAGE POWER
Embodied Energy	The total primary energy that is consumed for making the product.	162 MJ (1 W/s)
Operational Energy	The amount of the energy spent during system's lifetime.	6 W/s

they measured power consumption of a femtocell base station is idle, as the offered load is varied and as the datagram size is varied. In the experiment, they used one femtocell that supports up to four simultaneous end-user devices and it is connected to campus network. They also predicted the energy consumption of voice and FTP. In their prediction, they neglected that radio energy must be expended by the base station when making downlink transmission. Since in voice call, the downlink is active for a small period and active for the majority of the time during FTP download, they expected that radio power consumption should be higher than voice calls.

In [54], authors proposed power consumption model, it is based on a femtocell that consists of three interacting block: microprocessor, FPGA and radio frequency transmitter and power amplifier. They used energy efficiency, which is defined as power consumption needed to cover a certain area, in order to compare different technologies. They used ITU-R P.1238 propagation model and office scenario is assumed. They considered frequency, the floor penetration loss factor, the number of floors between base station and terminal and the distance power loss coefficient to calculate the range. Based on this model, they compared energy consumption for different bit rates and different technologies. Moreover, they used this model in a deployment tool that allows to design energy efficient femtocell networks by using genetic algorithm. Table 2.5 summarizes the aforementioned energy consumption models.

Table 2.5 Comparison of Different Energy Consumption Models

REFERENCE	MODEL TYPE	INPUT
[53]	Analytical	Datagram size (byte), offered load (Mbps), baseline power consumption when the femtocell is idle (Watts)
[54]	Experimental	Power consumption (Watts) of the microprocessor, the FPGA and the power amplifier (input power of antenna and the efficiency of the power amplifier which is the ratio of radio frequency output power to electrical input power)
[51]	Experimental	Static power consumption (idle mode power consumption of power amplifier, base band transceiver units, feeder network and cooling system) and dynamic power consumption (load on the base station and backhaul power consumption of base station)
[52]	Analytical (Log-normal distribution)	Radio frequency output power at maximum load, minimum load and in sleep mode and the dependency of the required input power on the traffic load

B. Energy-Efficient Deployments

Deployment strategy of 5G network is a very important design factor that has caught the attention of many researchers. Even though 5G technology is engineered to operate in different fields e.g. internet of things, it's the cellular phone applications that will be the first to truly embrace the full capabilities of 5G technology. Authors in [55] claim that massive MIMO (Multiple-Input Multiple-Output) has emerged as one leading technology enabler for the next generation mobile communications i.e. 5G. Most of the investigated MIMO scenarios consider preferably wide area outdoor deployments [56]. However, indoor users [57], generate most of the mobile traffic. In 5G, the main traffic volume of mobile services is occupied by high data rate services at indoors and at hotspots [58]. In this order to be more specific, we focus in the following on 5G Femtocells deployments.

Femtocells can be deployed in heterogeneous networks with different combinations of macrocells, microcells, picocells and femtocells and the reduction in total power consumption can be obtained from these schemes. The amount of energy efficiency depends on different variables. However, how to create this combination is also a challenging issue and there are some criteria to choose which cell to deploy. Although energy efficiency is not the first goal in some of the combinations modeled, energy efficiency is achieved. In general, cells are deployed depending on mobile user density, traffic and coverage in the network models. It is also important to provide same or better QoS when energy efficiency is subjected. In majority of the aforementioned studies, the network criteria mainly depend on the mobile user density, traffic and coverage. Mobile user density and traffic is considered to choose the correct cell size. Urban and rural areas are also important criteria in order to choose the cell. In the areas that have very weak signals can be supplemented with smaller cells such as femtocell so that coverage can be increased and there will be reduction in total power consumption.

There are different models in the literature to analysis energy efficiency [59] using heterogeneous network includes femtocell and macrocell. In [60], authors considered a network in which cells of different sizes have been deployed depending on mobile user density, traffic and coverage such that power consumption can be minimized without compromising with the QoS. They developed analytical models of power consumption in the proposed 5 different schemes and obtained the reduction in power consumption compared to macrocell network. In the first one, they used femtocell based network instead of macrocells in the area fully covered with femtocells and they obtained between 82.72% and 88.37% reduction in power consumption. In the second schema, they divided the area into three part as urban, suburban and rural. They considered mobile user density, mobile user traffic and required coverage and they covered urban areas by femtocells, suburban areas with macrocells and rural areas with portable femtocells. As a result of this simulation, they succeed between 78.53% and 80.19% reduction in power consumption. In the third schema, they allocated femtocells to densely populated urban area, picocells to sparsely populated urban areas, microcells to suburban areas and portable femtocells to rural areas and the reduction in power consumption rate was obtained as between 9.19% and 9.79%. In the fourth schema, they allocated microcells, picocells and femtocells to the border region and macrocell to remaining region. The reduction in power consumption is between 5.52% and 5.98% for this combination. The last one, femtocells are allocated between the boundaries of macrocell, where signal is not enough for a call. The result of the last model was between 1.94% and 2.66% reduction in power consumption and macrocell coverage.

In the analysis conducted by Bell Labs, scientists analyzed the efficiency of a hybrid network of both femtocell and macrocell, with open access femtocell, where all subscribers can connect to these femtocells like any base station [61]. The implementation area was 10 km by 10 km urban area of New Zealand and the population was about 200,000 people that means 65,000 homes and the 95% of population uses mobile equipment. They deployed varying number of femtocells that can serve up to 8 users in 100 m by 100 m area with 15W energy consumption. The reason of higher consumption rate compared to other analysis is that it is an open access model. They used continuously operated macrocells with 2.7kW energy consumption and femtocells deployed randomly in the area. The result of this analysis was depending on the network used: voice call and data connections. When femtocells mainly used for voice call, there was no big saving on total energy. On the other hand, when femtocells used for data connections, femtocells were able to reduce total energy consumption up to 60%. In this analysis, macrocells were used to ensure coverage and femtocell to offload capacity.

Another one was from Ofcom (the UK telecoms company) and Plextek as consultant [53]. Researchers there have analyzed two approaches; first, they deployed femtocells to 8 million households which is nearly 25% of the UK. Each femtocell consumed 7W every day and total annual energy consumption was about 490 GWh. In the second approach, they modelled macrocell network in order to provide indoor coverage, same with first approach and they would need 30,000 base stations in order to provide coverage. The model estimated that it takes 40 times more energy to deliver signal to indoor from macrocell, compared to femtocell. As a result of this analysis, in order to provide same coverage provided by femtocells, the total annual energy consumption of macrocell was 700 GWh per operator and 3500 Gwh for 5 operators. Thus the ratio of energy consumption in order to provide same indoor coverage in the UK was 7:1 over using macrocells

C. Power Consumption at Femtocells

The Femtocell access type sets the rules about who can connect to femtocell base station. It can be categorized in three type: open, closed and hybrid. In the literature, most of the studies about access type examines interference, QoS and handover issues. On the other hand, although there is no work directly compares the energy efficiency related the access type, energy consumption can be compared with examining different models in the literature. In open access, all available sources are shared between users and everyone can connect the network. They are mainly deployed in public places and there is no restriction to connect the network. Closed access, unlike open access, only closed subscriber group can connect the network but there can be different service levels between users and they are mainly used in small buildings. Hybrid access is combination of open and closed so that it allows particular outside users to access a femtocell. However, the conditions to connect the network outside the user group are defined by the operator and new entries to the system are requested by owner. These users (outside) can get only limited service, depends on the operator management. The comparison of the access types is given in Table 2.6.

Table 2.6 Comparison of Access Types

	OPEN ACCESS	CLOSED ACCESS	HYBRID ACCESS
Deployment	Public places	Residential deployment	Enterprise deployment
Number of handovers	High	Small	Medium
Owner preference	No	Yes	Yes
High user densities	Yes	No	No
QoS	Low	High	High
Interference between femtocell and macrocell	Increase	Decrease	Decrease

Although there is no work that compares the power consumption based on different access mode, it can be resulted as open access consumes more energy. In [61], power consumption of the femtocell is assumed as 15W although it is assumed as 6W in [50]. The reason of the high power consumption in [61] is the open access. Bit rate is another important factor that affects energy consumption of a femtocell. It can be defined as number of bits that are conveyed or processed per unit of time. As it reaches higher rates, the speed of connection becomes better since it is speed based measurement. Cellular operators try to increase the bitrates since it is important in market. On the other hand, bit rate has impact on energy consumption. In general, as bit rate increases, energy consumption also increases. Moreover, energy consumption for different bit rates is not same in different wireless technologies.

As we aforementioned, energy consumption of femtocell is not stable for different wireless technology standards such as WiMAX, HSPA and LTE which are rivals in the sector. WiMAX (Worldwide Interoperability for Microwave Access) can be used for transferring data across an ISP (Internet service provider) network, as a fixed wireless broadband Internet access, replacing satellite Internet service or as a mobile Internet access. HSPA (High Speed Packet Access) is another wireless technology standard, which is enhanced version of 3G. LTE (Long Term Evolution) is considered as 4G and provides better capacity and speed. They provide different bitrates and energy consumption of a femtocell is different for these technology standards. On the other hand, it is hard to say which one provides better energy efficiency since it changes with different bitrate ranges.

In [54], authors investigated energy efficiency of a femtocell base station and compared bit rates and for different wireless technologies, WiMAX, HSPA and LTE. Based on this model, they found that femtocell consumes nearly 10 W for a range between 9 to 130 m and WiMAX is the most energy efficient technology for bit rates more than 5 Mbps and LTE is the most energy efficient technology for bit rates between 2.8 and 5 Mbps. They used this model in a deployment tool that allows to design energy efficient femtocell networks by using genetic algorithm and they concluded WiMAX is the most energy efficient one for this scenario.

Network type also affects the energy consumption of femtocell. To illustrate, its power consumption is different for voice call and data transfer. Moreover, there are different data transfer protocols such as FTP and UDP. Datagram size and offered load have also impacts on the consumption. In [53], authors studied the effects of network type, datagram size and offered load. As the result of the analysis conducted by Bell Labs [61], the efficiency was depending on the network used: voice call and data connections. When femtocells mainly used for voice call, there was no big saving on total energy consumption. On the other hand, when femtocells used for data connections, femtocells were able to reduce total energy consumption up to 60%.

The last factor that has impact on femtocell power consumption is sleep mode. Since femtocell provides very small coverage compared to macrocell and very few users connect it, especially when it is deployed indoor, at most of the times it becomes idle. However, it consumes energy even if it serves no user. In this case, it is better to switch off the base station so this can be implemented with sleep mode. Sleep mode is energy efficient and feasible but the decision is also important. There are few works in the literature that analyzes the impact of sleep mode. In [54], authors examined sleep mode to reduce power consumption and it reduces the power consumption supporting up to 8 users and it led to 24% of power consumption in the network. Table 2.7 gives the summary of the section.

D. Smartphone/UE Power Consumption

One of the main idea of femtocell is to become closer to the user equipment (UE) which is a way to increase the system capacity and

Table 2.7 Factors Affecting the Power Consumption of a Femtocell BS

FACTOR	ENERGY CONSUMPTION
Access mode	Higher in open access, less in close access
Bit rate	Increases with higher bit rate
Wireless technology standards	Depends on bit rate range
Network type	Higher in data transfer
Sleep mode	Less energy consumed if sleep mode is an option

reduce energy consumption of both cellular network and UE [53]. However, since the coverage area of femtocell is not wide, the number of handover is very high in femtocell network. Moreover, UE uses most of its energy for handover process [62]. Another issue about handovers is that they decrease QoS and network capacity. Thus, handover decision algorithm is an important issue in femtocells. Although there are lots of studies about handover decision, only few of them depends on energy efficiency.

In [62], authors studied the energy efficiency of a femtocell per UE battery. They worked on the fact that there is a reduction in power consumption of UE under femtocell coverage, compared to macrocell connection and they proposed a handover decision algorithm that aims to reduce UE power consumption while maintaining QoS. The suggested algorithm enhances the strongest cell handout policy using an adaptive handout hysteresis margin. Although there is a need for increased LTE network signaling in the proposed algorithm, they derived power consumption and also interference. They compared the results with different algorithms in the literature and compared to a strongest cell based handout decision algorithm, the proposed algorithm reduced the UE energy consumption per bit by up to 85% with respect to femtocell deployment in LTE network.

VII. OPEN ISSUES

Our view of the core design problem in smartphones usage depends strongly on a clear understanding about their functionalities, and on required representations of: i) the environment they operate in, ii) the available resources (nodes, components, their costs and their accuracies), iii) the lifetime expectancy, and iv) the development in their operations and maintenance. The design process should further overcome the complications resulting from coupling application requirements and interfaces, and the underlying components and topology required for each. A significant challenge also lies in adequately representing the resources available in a given region of interest; those belonging to sensing nodes, such as transceivers, sensors and processors; and, those that exist under other systems. The most prominent sources of those that exist under other systems fall under the umbrella of industrial and municipal resources. They are deployed in abundance to collect information and control different operations and machinery.

Since user privacy is a concern for many users, researches design applications accordingly [27][32]. Many users do not install apps that collect unnecessary information thinking of privacy. Games, entertainment apps and shopping apps necessarily does not need permissions that can frustrate users. On the other hand, some application categories really need to get some private information from users. For example, navigation apps stop working when they cannot get location permission. Friend finding apps, health apps also needs some private information. Moreover, some app need to encrypt network communications in order to secure user related information [32]. Application categories that take encryption as a primary design factor can be banking apps, social media and messaging. On the other hand, some application categories can consider encryption as a secondary design factor like alarm apps, translation apps, etc. Further research shall pitfalls of the aforementioned security/privacy issues in addition to the single application-specific design, which stands as a major hindrance in realizing public sensing in smartphone usage. It shall allow for great flexibility in design and an expandable and/or upgradeable set of functionalities. It shall allow for the reutilization of existing sensing resources and infrastructures across applications, resulting in extended functional operations and increased return on investment.

Moreover, this survey shows that there are variety of issues on the energy efficiency of femtocell networks needing to be investigated in the future, including:

- Energy metrics and energy consumption models considering not only femtocells but also femtocell and macrocell heterogeneous networks.

- Energy consumption models, considering more details in order to make more accurate estimations.
- Comparison of energy efficiency of femtocells when they are deployed with different access types.
- The optimized deployment of cellular networks in huge area, including one macrocell and many overlapping femtocells.
- Femtocell handover decision while considering the network energy-efficiency.
- The effect of femtocell on UE energy consumption, considering different parameters such as the access type.
- Evaluating the efficiency of femtocell in terms of energy consumption as a communication technology in IoT applications and comparison with other communication technologies.
- The effect of interference problem on femtocell energy consumption.
- All the above mentioned issues are still open research problems that necessitate further investigations in order to realize the vision of 5G enabled devices and smart spaces.

VIII. CONCLUDING REMARKS

In this research, we believe that the idea of monitoring certain types of smartphone activities will minimize the users concerns about accessibility and tracking in smart environments. The main drawback of the existing methods is their intrusive and disruptive nature, due to the fact that users may need to be interrupted from what they are doing in order to provide a piece of information such as a password. Although some methods are less disruptive (such as face recognition), they are still intrusive as the use of a camera or other sensor types means highly personalized information are collected and stored by the system to be used in the verification process of any context aware system [11]. This also raises a question with regards to users' perception and tolerance to such methods as people are concerned how, where else, and by whom their information is being used. Similarly, context aware systems need to collect a variety of information about the user's current status and activities, some of which may be regarded as personal. New proposals are required to address this issue by only collecting information related to the user's access to resources, which has to be collected in any case. Collecting such information does not require the users to perform additional disruptive activities in the process of verifying their identity. As the information needed by the framework is collected anyway and no additional information is required (collected) the authentication decisions can be made more quickly. Furthermore, energy efficiency of femtocell networks become more important with the increasing deployment of huge numbers of femtocells. Moreover, it can be used in order to manage energy efficiency when it is used in smart grids. In this article, we mentioned energy efficiency of femtocell networks in IoT, considering energy metrics, energy consumption models, energy efficiency schemes for deployment, factors that affects energy consumption of femtocell networks and energy efficiency of UE under femtocell coverage. Energy efficiency of femtocell in IoT still needs to be investigated.

REFERENCES

[1] Gupta, Akhil, and Rakesh Kumar Jha. "A survey of 5G network: Architecture and emerging technologies." IEEE access 3 (2015): 1206–1232.

[2] Uckelmann, Dieter, Mark Harrison, and Florian Michahelles. "An architectural approach towards the future internet of things." Architecting the internet of things. Springer Berlin Heidelberg, 2011. 1–24.

[3] LM Ericsson, "More than 50 Billion Connected Devices," February 2011. www.ericsson.com/res/docs/whitepapers/wp-50-billions.pdf.

[4] A. P. Castellani, N. Bui, P. Casari, M. Rossi, Z. Shelby and M. Zorzi, "Architecture and protocols for the internet of things: A case study," *IEEE PERCOM*, 2010, pp. 678–683.

[5] L. Atzori, et. al., "The Internet of Things: A survey", *Computer Networks*, Vol. 54, No. 15, 2010, pp. 2787–2805.

[6] G. Barrenetxea, F. Ingelrest, G. Schaefer and M. Vetterli, "The hitchhiker's guide to successful wireless sensor network deployments", *In Proc. of the 6th ACM conference on Embedded network sensor systems (SenSys)*. ACM, pp. 43–56, 2008.

[7] F. Al-Turjman, H. Hassanein, W. Alsalih, and M. Ibnkahla, "Optimized Relay Placement for Wireless Sensor Networks Federation in Environmental Applications", *Wireless Comm. & Mobile Comp. Journal*, vol. 11, no. 12, pp. 1677–1688, Dec. 2011.

[8] F. Al-Turjman, H. Hassanein, and M. Ibnkahla, "Optimized Relay Repositioning for Wireless Sensor Networks Applied in Environmental Applications", *In Proc. of the Inter. W. Comm. and Mob, Comp, conf.*, Istanbul, Turkey, 2011, pp. 1860–1864.

[9] F. Al-Turjman and H. Hassanein, "Enhanced data delivery framework for dynamic Information-Centric Networks", *In Proc. of the Local Comp. Net. (LCN)*, Sydney, AU, 2013, pp. 831–838.

[10] G. Anastasi, M. Conti, M. Francesco, A. Passarella, "Energy conservation in wireless sensor networks: A survey", *Ad Hoc Networks*, Vol. 7, Iss. 3, pp. 537–568, May 2009.

[11] F. Al-Turjman, "Cognition in Information-Centric Sensor Networks for IoT Applications: An Overview", *Springer Annals of Telecommunications Journal*, pp. 1–16, 2016.

[12] A. Salhieh, J. Weinmann, M. Kochhal, L. Schwiebert, Power Efficient Topologies for Wireless Sensor Networks, International Conference on Parallel Processing (ICPP), 2001.

[13] M. Biglarbegian and F. Al-Turjman, "Path Planning for Data Collectors in Precision Agriculture WSNs", *In Proc. of the International Wireless Communications and Mobile Computing Conference (IWCMC)*, Nicosia, Cyprus, 2014, pp. 483–487.

[14] R. Kuntz, A. Gallais and T Noel, Medium access control facing the reality of WSN deployments. SIGCOMM Comput. Commun. Rev., Vol 39, Iss. 3 (June 2009), pp. 22–27.

[15] N. Lane, E. Miluzzo, L. Hong, D. Peebles, T. Choudhury, A. Campbell, "A survey of mobile phone sensing," *IEEE Comm. Magazine*, vol. 48, no.9, pp.140–150, Sept. 2010.

[16] F. M. Al-Turjman, H. S. Hassanein, M. Ibnkahla, "Efficient deployment of wireless sensor networks targeting environment monitoring applications, *Computer Communications*, vol. 36, no. 2, pp. 135–148, 2013.

[17] Rahmati, A., & Zhong, L. (2009). Human–battery interaction on mobile phones. *Pervasive and Mobile Computing*, vol. 5, no. 5, pp. 465–477.

[18] Wagner, D. T., Rice, A., & Beresford, A. R. "Device analyzer: Understanding smartphone usage". *In Mobile and Ubiquitous Systems: Computing, Networking, and Services*, pp. 195–208.

[19] Ferdous, R., Osmani, V., & Mayora, O. "Smartphone app usage as a predictor of perceived stress levels at workplace". *In Proc. of the 9th Inter. Conf. on Pervasive Computing Technologies for Healthcare*, pp. 225–228, 2015.

[20] Do, T. M. T., Blom, J., & Gatica-Perez, D., "Smartphone usage in the wild: a large-scale analysis of applications and context", *In Proceedings of the 13th ACM International Conference on Multimodal Interfaces*, pp. 353–360, Nov. 2011.

[21] F. Al-Turjman, "Mobile Couriers' Selection for the Smart-grid in Smart cities' Pervasive Sensing", *Elsevier Future Generation Computer Systems*, 2017. DOI: 10.1016/j.future.2017.09.033.

[22] Altmann, J. and Sampath, R., 2006, UNIQuE: A User-Centric Framework for Network Identity Management, Network Operations and Management Symposium, NOMS 3-7 April 2006. 10th IEEE/IFIP, 495–506. doi: 10.1109/NOMS.2006.1687578.

[23] Koshutanski, H., Ion, M. and Telesca, L., 2007, Distributed Identity Management Model for Digital Ecosystems, *The International Conference on Emerging Security Information, Systems, and Technologies*. Oct. 2007, pp. 132–138.

[24] F. Al-Turjman, and M. Gunay, "CAR Approach for the Internet of Things (IoT)", *IEEE Canadian Journal of Electrical and Computer Engineering*, vol. 39, no. 1, pp. 11–18, Winter, 2016.

[25] Li, H., Lu, X., Liu, X., Xie, T., Bian, K., Lin, F. X., & Feng, F. (2015, October). Characterizing Smartphone Usage Patterns from Millions of Android Users. *In Proc, of the 2015 ACM Conference on Internet Measurement Conference* (pp. 459–472). ACM.

[26] K. Elgazzar, A. Ejaz and H. Hassanein, "AppaaS: offering mobile applications as a cloud service", *Journal of Internet Services and Applications*, 2013.

[27] Rahmati, A., & Zhong, L., "Studying smartphone usage: Lessons from a four-month field study", *IEEE Transactions on Mobile Computing*, vol. 12, no.7, pp. 1417–1427.

[28] Liao, Z. X., Li, S. C., Peng, W. C., Yu, P. S., & Liu, T. C. (2013, December). On the feature discovery for app usage prediction in smartphones, *IEEE 13th International Conference on Data Mining (ICDM)*, (pp. 1127–1132). 2013.

[29] H. Falaki, R. Mahajan, R. Mahajan, D. Lymberopoulos, R. Govindan, D. Estrin, "Diversity in Smartphone Usage".

[30] Lu, E. H. C., Lin, Y. W., & Ciou, J. B. "Mining mobile application sequential patterns for usage prediction". *In IEEE International Conference on Granular Computing (GrC)*, pp. 185–190, 2014.

[31] Montoliu, R., & Gatica-Perez, D., "Discovering human places of interest from multimodal mobile phone data", *In Proceedings of the 9th International Conference on Mobile and Ubiquitous Multimedia*, pp. 12, December, 2010.

[32] Samsung S II Available online.: https://www.anandtech.com/show/4686/samsung-galaxy-s-2-international-review-the-best-redefined/13.

[33] Zheng, Y., Zhang, L., Ma, Z., Xie, X., & Ma, W. Y. (2011). Recommending friends and locations based on individual location history. *ACM Transactions on the Web (TWEB)*, 5(1), 5.

[34] F. M. Al-Turjman, H. S. Hassanein, M. Ibnkahla "Quantifying connectivity in wireless sensor networks with grid-based deployments", *J. Network and Computer Applications*, vol. 36, no. 1, pp. 368–377, 2013.

[35] _____,"Testing Strategies and Tactics for Mobile Applications, Keynote White Paper" (PDF). Keynote.com. Retrieved 2012–05-02.ASD.

[36] Saeed, R. A. (Ed.). (2012). Femtocell Communications and Technologies: Business Opportunities and Deployment Challenges: Business Opportunities and Deployment Challenges. IGI Global.

[37] Smart grids and meters - Energy - European Commission. (n.d.). Retrieved November 21, 2016, from http://ec.europa.eu/energy/en/topics/markets-and-consumers/smart-grids-and-meters.

[38] V. Chandrasekhar, J. G. Andrews, and A. Gatherer, "Femtocell Networks: A Survey," *IEEE Commun. Mag.*, vol. 46, no. 9, 2008, pp. 59–67.

[39] National Intelligence Council, "Disruptive civil technologies—six technologies with potential impacts on us interests out to 2025," Conference Report CR 2008-07, 2008.

[40] Chowdhury, M. Z., Lee, S. Q., Ru, B. H., Park, N., & Jang, Y. M. (2011, September). Service quality improvement of mobile users in vehicular environment by mobile femtocell network deployment. In ICTC 2011 (pp. 194–198). IEEE.

[41] S.R. Hall, A.W. Jeffries, S.E. Avis, D.D.N. Bevan, Performance of open access femtocells in 4G macrocellular networks, in: The Wireless World Research Forum 20 (WWRF 20), Ottawa, Canada, 2008.

[42] M. Domingues, and A. Radwan, "Optical Fiber Sensors for IoT and Smart Devices," *Springer Briefs in Electrical and Computer Engineering*, March 2017, DOI: 10.1007/978-3-319-47349-9.

[43] A. Radwan, K. Huq, S. Mumtaz, K. F. Tsang, J. Rodriguez, "Low-cost On-demand C-RAN based Mobile Small-Cells," IEEE Access, vol. 4, pp. 2331–2339, May 2016, DOI: 10.1109/ACCESS.2016.2563518.

[44] A. D. Domenico, R. Gupta, and E. Calvanese, "Dynamic Traffic Management for Green Open Access Femtocell Networks", Proc. IEEE VTC-Spring, 2012.

[45] Fan, Z., Kulkarni, P., Gormus, S., Efthymiou, C., Kalogridis, G., Sooriyabandara, & Chin, W. H. (2013). Smart grid communications: Overview of research challenges, solutions, and standardization activities. IEEE Communications Surveys & Tutorials, 15(1), 21–38.

[46] Hindia, M. N., Rahman, T. A., Ojukwu, H., Hanafi, E. B., & Fattouh, A. "Enabling Remote Health-Caring Utilizing IoT Concept over LTE-Femtocell Networks". PloS, vol. 11, no. 5, 2016.

[47] Cho, T. H., & Jeon, G. M. "A method for detecting man-in-the-middle attacks using time synchronization one-time password in interlock protocol based internet of things". *Journal of Applied and Physical Sciences*, vol. 2, no. 2, pp. 37–41, 2016.

[48] F. Al-Turjman, H. Hassanein, S. Oteafy, and W. Alsalih, "Towards augmenting federated wireless sensor networks in forestry applications", Springer: Personal and Ubiquitous Computing Journal, vol. 17, no. 5, pp. 1025–1034, June, 2013.

[49] ABI Research, High Inventory and Low Burn Rate Stalls Femtocell Market in 2012 (July 5, 2012 November 13, 2012.).

[50] J. Zhang et al., "A Novel Power Control Scheme for Femtocell in Heterogeneous Networks," Proc. IEEE CCNC, 2012.

[51] Y. Hou and D. I. Laurenson, "Energy Efficiency of High QoS Heterogeneous Wireless Communication Network," IEEE VTC-Fall, 2010

[52] M. W. Arshad, A. Vastberg, and T. Edler, "Energy Efficiency Gains Through Traffic Offloading and Traffic Expansion in Joint Macro Pico Deployment," Proc. IEEE WCNC, 2012.

[53] Hashmi, Z. H. (2013, May 04). Adaptive and Efficient Resource Management for Emerging Wireless Networks. Electronic Theses and Dissertations (ETDs) 2008. doi:10.14288/1.0073687.

[54] Riggio, R., & Leith, D. J. (2012). A measurement-based model of energy consumption in femtocells. 2012 IFIP Wireless Days. doi:10.1109/wd.2012.6402872.

[55] Panzner, Berthold, et al. "Deployment and implementation strategies for massive MIMO in 5G." Globecom Workshops (GC Wkshps), 2014. IEEE, 2014.

[56] E. G. Larsson, O. Edfors, F. Tufvesson, and T. L. Marzetta, "Massive mimo for next generation wireless systems," IEEE Commun. Mag., vol. 52, no. 2, pp. 186–195, Feb. 2014.

[57] J. Zhang and G. de la Roche, Eds., Femtocells: Technologies and Deployment. Wiley, 2013.

[58] Chen, Shanzhi, and Jian Zhao. "The requirements, challenges, and technologies for 5G of terrestrial mobile telecommunication." *IEEE Communications Magazine*, vol. 52, no. 5, pp. 36–43, 2014.

[59] F. Al-Turjman, "Energy–aware Data Delivery Framework for Safety-Oriented Mobile IoT", *IEEE Sensors Journal*, 2017. DOI: 10.1109/JSEN.2017.2761396.

[60] Deruyck, M., Vulder, D. D., Joseph, W., & Martens, L. (2012). Modelling the power consumption in femtocell networks. IEEE Wireless Communications and Networking Conference Workshops (WCNCW).

[61] Feng, Z., & Yuexia, Z. (2011). Study on smart grid communications system based on new generation wireless technology. International Conference on Electronics, Communications and Control (ICECC).

[62] Mukherjee, A., Bhattacherjee, S., Pal, S., & De, D. (2013). Femtocell based green power consumption methods for mobile network. Computer Networks, 57(1), 162–178 doi:10.1016/j.comnet.2012.09.007.

3

SECURITY IN MOBILE IoT SPACES

FADI AL-TURJMAN AND CHADI ALTRJMAN

Contents

Abstract

With the continuous technological development, it has become certain that VANETs will play an important part in our emerging smart environments. It is becoming more and more difficult to assure the safe functioning of smart systems as they are becoming increasingly susceptible to malfunctions and external attacks, especially in dynamic environments such as the vehicular system. In this chapter, we have established a brief overview of the current vehicular ad hoc network (VANET) model and the constraints associated with it for security access, followed by a cataloguing in terms of the wireless messages attackers and attacks in addition to their appropriate defence strategies. Open research issues and challenges have been pointed out as well for further investigations and enhancements.

Keywords

IoT, Security, VANETs, IoT-based Attacks.

I. INTRODUCTION

Considering the rate at which VANETs have been developing and the benefits expected from vehicular interconnections with millions of vehicles in traffic worldwide, it is most likely that they will be the most concrete and common usage of mobile ad-hoc networks (MANETs) [1][2]. With the future possibility of seamless connectivity expected from 5G with high backward compatibility and highly integrative design [3], proper implementation and dissemination of onboard units (OBUs), geolocational devices such as GPS or GLONASS (Global Navigation Satellite System) receivers, combined with the ever-increasing communicational potential, create significant opportunities for economical solutions. However, it also raises critical research challenges and important ethical questions [4][5].

The increasing connectivity between numerous elements has made efficient security protocols and applications a prime subject of study. With the massive amount of data that can be produced by modern vehicles that should be treated in a timely manner, efficiency is a paramount in a system where an error or an attack can result in the loss of human life. One of the primary constraints of VANETs compared to most systems is to make sure life-critical information is delivered on time while establishing the liability of the users and simultaneously protecting the privacy of the drivers to its fullest extent. Balancing both the constraints of this medium with effective security requires us to fulfil numerous parameters without exceeding the need of one and vice versa. VANETs by their nature have some advantages and disadvantages in comparison to MANETs and IoT networks. Detailed analysis of this properties needs to be established to develop flexible and modular solutions [6].

The type of communication transmitted in the network has different constraints and urgency in accordance with the content of the communicated information. The way security is applied need to be in proportion to the confidentiality required for that specific message. This creates a need to classify different levels of communication before even broadcasting them. Proper privacy enforcement requires cooperation between the governmental institutions and private companies. The ethical and technical applications, both theoretical and practical, needed for effective treatment of information while conserving enough privacy needs in-depth discussions. Understanding the nature of potential attacks and disturbances both intentional and accidental has made it a requirement for researchers on the topic to classify them according to different existing criteria.

In the first section, we will detail our system model by defining what are the characteristics, both positive and negative, of VANET in the IoT paradiam and smart spaces in general. We will also discuss the message constraints and the message types that will be delivered across the network to identify critical and constrained messages. The second section contains a general attacker classification, attacks discussed in the literature and potential defence methods that have been applied in particular to them. The third part contains the overall classification of defence methods and the presentation of some of the techniques that fall under that particular umbrella. The last part contains some potential areas of research and development that are found to be lacking or understudied in the security software protocols of VANETs and that could lead to interesting results in the future.

II. SYSTEM MODEL

We must first define the standard features of VANETs before discussing any theoretical or practical application of security in the networks. It is also important to understand the constraints shared by a majority of VANETs, either physical or architectural. We will also define the constraint that security models and broadcasted messages must fulfil in such a network.

A. Network Model

The nodes present in VANETs can be either a mobile vehicle or a roadside unit (RSU). Vehicles can be either private or public while RSUs can belong to governmental authorities or private service providers such as telecommunication providers but in some cases, a parked vehicle can be considered a RSU. The large scale of the VANETs is another main feature that creates the need to set them apart from other IoT networks and MANET. With countless nodes distributed on a broad scope, VANETs are most likely to become the largest actual application of a MANET. However, due to its application and cluster based topography, communication in the network at large will be mainly local, thus allowing for the partition of the network and easing it scalability [7].

The greater majority of the network nodes consists of vehicles, thus the network dynamics are characterised by high speeds, quasi-permanent mobility and short connection times between members. One aspect of network dynamics that makes it slightly less chaotic is the predictable vehicle trajectories which are mostly well defined by the roads, which offers some advantages for the dissemination of messages and the reduction of randomness compared to most MANET while creating the disadvantages of an increase in potential breaches leading to disturbances in the provided services.

Another distinct and major advantage of VANETs compared to more traditional ad hoc networks is the nonnegligible computational and power resources the vehicles can easily provide. A standard "smart" vehicle in a VANET hosts hundreds or even thousands of microprocessors, an Event Data Recorder (EDR) that can be put to use for incident reconstruction, and a geolocational system that can indicate position and time.

The primary function of VANETs should be operational without dependence on heavy infrastructure presence nor the majority of other "smart" cars on the road as the two will keep co-existing for the following decades.

B. VANET Application

We have divided the potential uses of VANETs into two different categories depending on their level of importance from a security point of view. They have a vast range and some are currently in use while others are more futuristic and are only discussed in theory.

The first one is the safety-related applications, with a concrete example being cooperative driving and collision avoidance. The defining characteristic of this type of application is its relevance to emergency situations where the presence of a service may prevent or help cope with life-threatening accidents. The security of this category, be it physical or electronic, is a must as the proper operations of any of these functions should be guaranteed even in the presence of disturbances, be it due to errors or attacks.

The other category includes the not urgent applications, including infotainment, location-dependent services, traffic optimization and payment services. It cannot be denied that security is without a doubt a requirement in this category, especially when considering the cases where monetary transactions are involved. The examples of attacks and defences presented further will mostly focus its attention on the security aspects of safety-related applications as they are often more concerning when looking specifically from the automotive domains perspective and because they usually raise the most challenging problems, possess harsher constraints and carry greater risks. Some attention will still be given to other applications, especially concerning potential identity theft and economic damage.

C. *Message Requirements*

A security model developed for use in VANETs should by default be able to satisfy the following conditions and requirements for its safety messaging:

- **Privacy:** People have become more aware and concerned about tracking and invasive surveillance technologies. The privacy of users and members against unauthorized observers has to be guaranteed. This is a chief concern as the development of VANETs will follow customer demand as car companies will most likely not innovate if it cannot sell its ideas.
- **Authentication:** reactions given by vehicles to events should be based on legitimate messages. Therefore proper protocols of authentication should be employed by the senders of these messages be it by public keys given out by the government or local network keys.
- **Time Critical:** Considering the very high speeds expected in a typical VANET, stringent constraints should be expected when dealing with time-sensitive data as it might not leave any room for mistakes and could have disastrous results if not met properly.
- **Data Plausibility:** the legitimacy of transmitted messages also includes the evaluation of their consistency with similar ones, as the legitimacy of the broadcaster can be assured while

the contents of the message contain erroneous data. The way that the plausibility is confirmed will firmly depend on the type of data transferred.

- **Availability:** Even when we assume the existence of a robust communication channel, some attacks and malfunctions can weaken and bring down the network be it by sheer volume or finding flaws in the system. Therefore, it is paramount that availability of services should be also supported by alternative means. This can be a V2V or V2I solution but a backup protocol should be in place.
- **Non-repudiation:** Individuals and vehicles causing accidents need to be reliably identified while the sender should not be able to pick and choose which message to broadcast or deny the transmission of certain messages. This leads us to the last point which might be the hardest to address in this hyper-connected world.

D. Message Constraints

We mainly concern ourselves with applications where identity theft leading to economic or social damage is a possibility in smart spaces. Using this perspective, we can classify the safety messages into four distinct classes by weighting their properties related to privacy and real-time constraints.

- **Traffic information:** disseminates traffic conditions that might arise from natural or unnatural causes in a given region and thus affect public safety only indirectly; hence they are often not time-dependent but require general privacy and legitimacy.
- **General safety:** employed by public safety applications such as cooperative driving and collision avoidance and must satisfy strict time constraints. The fact that "smart" and "stupid" vehicles coexist in the system simultaneously and identifying them in time will be essential for VANETs to work properly.
- **Identity-Economic messages:** are not counted as critical yet are becoming an integral part of VANET communications as automatic toll payment, and wireless payment options become more prolific. Malicious individuals might intercept payment information and use them without the knowledge of the user.

Stolen identity, be it an individuals or cars, can create liability and security-related issues.

- **Liability:** distinguished from the previous class as they are exchanged in a situation where liability plays a part such as speeding, collisions, accidents or unusual events. Before the broadcast, the extent of the liability held by the originator of the message should be determined and reveal his identity to the authorities should it prove necessary. This is also one of the problems in effective integration of VANETs and market penetration. Selling vehicles that might report their owner to the law is a tough sell. It is an ethical question that must be solved and answered as to when the liability messages should be broadcasted to the authorities.

E. Trust and Privacy

The primary element in any security system is trust and privacy. This is particularly true and critical in VANETs due to the high liability expected from safety and security applications and consequently the members running them. With a significant number of independent nodes in the network and the presence of the human factor, it is without a doubt highly probable that misbehaviour will happen. In a connected world, users are increasingly concerned about their privacy and drivers are by no means an exception. This is especially problematic as the lack of privacy and the potential tracking functionality inherent in VANETs may result in fines on the drivers, leading in turn to a mistrust by the users towards VANETs. Due to the previous point, we should likely assume a low amount of trust in members and service providers. Beside the two main members, the drivers and service providers, it is without a doubt certain that there will be, and should be, a considerable presence of governmental authorities in these networks.

III. ATTACKS IN VANETs

As it is not realistic to envision and name all the potential attack avenues that can be mounted in the future on VANETs, we will strive to provide a general classification of both the possible forms taken by the perpetrators and the type of attack identified in the literature. We concern ourselves with VANETs; we will only consider and take into account the attacks that can be perpetrated against messages rather than directly on the vehicles, as the physical security of vehicles OBUs are outside of our scope and would require a whole other survey.

A. Attacker Classification

Understanding the nature of the attacker is important for classifying the types of attacks that VANETs might be subjected to. This classification is straightforward and is commonly accepted amongst researchers and engineers dealing with the subject.

- **Active/Passive:** active attackers generate packets or signals, meanwhile a passive attacker contents himself with eavesdropping in on the wireless channel.
- **Insider/Outsider:** an authenticated member of the network who can broadcast and receive messages from other members is an insider. It has access to a certified public key and can more easily navigate the network protocols for mounting attacks. Meanwhile, the outsider is considered as a foreign object by the network members and as an intruder. Hence, it is severely limited in his interactions with security protocols.
- **Local/Global:** an attacker can be limited in scope, even if he controls several nodes, which makes him local and limiting his impact at large. The extended attacker can control several entities that are scattered across the network, thus allowing him to be active on a larger scope. The differentiation between local and extended is not easy to make and depends heavily on the size and range of the VANETs in question. This distinction can be especially important in tracking privacy-violating activities and potential suspects.
- **Malicious/Rational:** a malicious attacker is not in search of any kind of personal benefits from the attacks he perpetrates and aims to harm the members or the integrity of the network and create havoc. The main issue is the fact he could employ any means disregarding corresponding costs and consequences making him unpredictable and potentially extremely dangerous.

On the contrary, an attacker that seeks personal profit is rational and will not overextend his resources for any intangible gain. Hence, this makes him more predictable regarding the attack means and the attack targets that he might choose to assault.

B. *Attacks*

The following examples are the types of attacks discussed in the literature. Some of them are simple to set-up and might even be unintentional or just natural misbehaviours in the network rather than the result of a conscious effort to disrupt the active system. We should also assume that as the complexity of the attack method increases the skill of the perpetrator increases accordingly. Attacks that deal mainly with tampered hardware and OBU are not part of the scope of this review [8]-[16].

- **Masquerade:** the member is actively masking its own identity to appear like another vehicle by using false identities, such as public keys. This technique is usually employed in conjunction with other types of attacks.
- **Location Tracking:** the observer can monitor the trajectories of selected members and can use this information for a range of purposes, both malicious and mundane. It can also potentially leverage on the RSU or the vehicles who are around its main target. It is more likely than not malicious and requires some preparation, such as disseminating a virus in said network or some prior physical access.
- **Cheating with sensor information:** attackers alter their perceived location and speed to escape liability. It can be used notably in the case of an accident. In the worst case, colluding attackers can clone each other, but this would require retrieving the security material.
- **False information:** transmits erroneous information and data in the network, which might affect the behaviour of other drivers. It can be both intentional and unintentional.
- **Denial of Service (DoS):** an attacker can bring down VANETs, jam signals or may even cause an accident by using malicious nodes to forge a significant number of bogus identities, such as IP addresses, with the final objective of

disrupting the proper functioning of data and information transfer between two fast-moving members. An example would be jamming the communications channels. It is the most likely of basic attacks perpetrated by a malicious attacker. Preventing DoS attacks has been a research topic of computer sciences and an extensive amount material is available [17]-[19] that could be applied to VANETs such as the IP-CHOCK scheme proposed in [20]. The IP-CHOCK approach is capable of locating suspicious nodes without requiring any kind of special hardware support or additional secret information exchange. Another approach is the Attacked Packet Detection Algorithm (APDA) presented in [21] where it is employed to detect and identify the DoS attacks before engaging in resource-intensive verification time in an effort to increase security and minimize overhead delay.

- **GPS Spoofing:** GPS satellites, or their equivalent, maintains a locational table with the spatial locations and identifiers of vehicles in the network. Attackers may produce misleading and false readings in the positioning system with the purpose to deceive vehicles, leading them to assume that they are actually in a different spot. It is relatively easy to dupe any amount of members as discussed in [22] with some restrictions. It is also quite possible to use a GPS satellite simulator to generate and broadcast signals that are stronger than those generated by the actual satellite system, leading the receivers to prefer it to the actual satellite.

- **Wormhole:** traditionally this is accomplished by tunnelling packets between two remote members of a network [23]. In a VANET, the perpetrator should control at least two nodes separate from each other and with a very high-speed connection between them to tunnel packets from one location to broadcast them in another. This could be accomplished with pre-established RSUs or by using mobile technologies such as 4G, or 5G in the near future. Wormholes allow the attacker to spread misleading but properly signed messages at the destination area. A way to protect a network from wormhole attacks is the TESLA with Instant Key disclosure (TIK) [23]. It is a method using packet leashes that calculates the divergence in the allowed theoretical travel distance and the actual travel distance of the transmitted packet to identify and spot any signs of tampering on the anomaly that might point to a potential attack. Another potential solution discussed in the literature is the AODV routing protocol where hop-by-hop efficient authentication protocol (HEAP) [24] is a useful approach that allows us to notice wormhole attacks.

- **Physical/Electronic Tunnel:** when a GPS signal disappears in a tunnel, it can be exploited by using the temporary loss of connection to the system to inject falsified data and positioning information. Once the vehicle leaves the tunnel, it will assume that this is his actual position and will tread as such before it receives an updated position from the satellite. A non-physical tunnel could be created using proper jamming which broadens the potential applications of such an attack.

- **Man-in-the-Middle:** malicious vehicles eavesdrop on the communication between vehicles and inject false information or distort messages between vehicles [14][15]. Reasonable solutions are strong cryptography, secure authentication and data integrity verifications.

- **Black Hole:** data packets get lost while crossing through a Black Hole, in effect a member that has some nodes or no node that refuse to broadcast or forward data packets to the next hop. Preventing black hole attacks is generally achieved by making use of the redundant paths kept between the sender and the destination of the message, although this has the unfortunate side effect of adding to the network complexity. Another potential way to defend against Black Holes is the use of an information carrying sequence number in the message header. The receiver can then potentially figure out the absence of a packet in the case of any discrepancy or loss, identifying the situation as suspicious [15].

- **Timing:** data transmission at the right time from one node to another is essential in achieving data security and integrity. An attack is labelled as Timing whenever a malicious vehicle receives a time-critical emergency message and do not forward it to their neighbouring members on time by adding some additional timeslots to the original in an effort to create an artificial delay in transmission and reaction. Thus, neighbouring vehicles to the attackers may receive the message outside the scope of the time constraint, rendering it moot.

- **Malware and Spam:** attacks, like spam and viruses, can lead to severe disruptions in VANETs operations. They are typically the work of malicious insiders rather than outsiders who have access to OBUs of vehicles and RSUs when they are performing software updates. They can potentially result in an increase in transmission latency, which can be lessened by using a centralised management. Proper maintenance of infrastructure and a centralized administration should be employed to prevent any such attack to the VANETs, the security of the OBUs of the mobile members are not in our scope.

- **Sybil:** The perpetrator creates multiple identities in an effort to simulate multiple nodes [25][26]. Each of this nodes can transmit messages, even with multiple differing identities. Thus, other vehicles are led to think that there are more vehicles in the network that there actually is. This type of attack is potentially extremely hazardous in the constraint heavy VANETs since a vehicle can also potentially claim to be in a different spot at the same time causing substantial security risks and causing chaos in the network. They are traditionally detected through the use of resource testing [8][12], but this approach assumes that all entities are bound to be limited to some resources. Computational PUZZLES are used in [8] to assess computational resource usages of each node. However, this technique cannot be said to be appropriate for use in Sybil attack detection in VANETs [12] as an attacker node can have potentially more computational power than an ordinary node encountered in other MANET and IoT systems. Instead of computational testing, the use of radio resource testing [12] can be used to identify the attack as the attacker node will most likely use more broadcast resource than standard nodes as it will broadcast like multiple members simultaneously. The use of public key cryptography [27] can prevent and eliminate the risk of Sybil attacks where the vehicles are authenticated using the public keys distributed by authorities. Key revocation can be another possible approach that may reduce the influence and impact of Sibyl attacks identified in wireless sensor networks [28][29] using a predefined model of propagation. Using this model we can measure the true distance of a node through the received signal strength indicator (RSSI) method, where the differences of the signal strength of the transmitted and received messages are compared while signals are matched with the nodes claimed spatial location. If the claimed spot by the node does not correspond to the defined distance limit set from the evaluated location for that model of propagation, this node is a potential Sybil attacker and should be treated as such.

- **Impersonation Attack:** During a V2V communications, a member can broadcast the security messages as if it was the origin to other vehicles which can potentially have an impact on the behaviour of the traffic control system and the other members of the VANET. This a case of an impersonation attack, a malicious vehicle transmits a message on behalf of another member to cause accidents, traffic jams, create chaos, or other security attacks while masquerading. The SPECS scheme, proposed in [31] and detailed in [32], ensures the privacy and security issues of V2V communications by detecting the impersonation attacks. This is accomplished with the use of an Identity-based batch Verification protocol which suffers from this type of attack and cannot fulfil its privacy requirements, leading to the discovery of the misbehaving member. To protect the identity of each vehicle, it uses a pseudo-identification code and a shared secret key between the RSU and vehicles present in the network. This type of attack is the reason the distribution of identity keys by the authorities is paramount for identifying the origin of the broadcasted messages.

- **Illusion Attack:** When an attacker broadcasts the traffic warning messages that do not correspond to the current road condition, producing an illusion to the members in their neighbourhood. The propagation of the phantasm mainly depends on the drivers' responses, which can lead to car accidents, traffic jams and a decrease in the overall health and performance of the VANET. Classic message control approaches cannot protect networks against illusion attacks due to the explicit control the adversary exerts on the sensors of its vehicle, misleading its inputs in an effort to create and broadcast false information. The Plausibility Validation Network (PVN), proposed in [11], is a potential model to secure VANETs against this type of assaults. PVN collects raw sensor's data, verifies whether the collected data seems realistic and plausible. It requires two distinct types of inputs: data transmitted from antennas and data sensed by sensors. They are then categorised by employing a header placed on the transmitted data. PVN possess a rule database and checking module for transmitting data which allows it to check the authenticity of the inputs and takes the necessary actions accordingly if found to be illusionary. A transmission is considered trustworthy if it can pass all verifications processes successfully. If it fails, it is treated as an illusion and dropped by default. PVN can potentially cooperate with different types of cryptographic methods to be able to defend against more various forms of attacks.

- **Purposeful/Intentional Attack:** the attacks perpetrated by insiders are problematic to prevent and defend against properly as they are already an authenticated member of the network and most likely trusted to perform V2V and V2I communications with their neighbours, making it difficult to catch. In VANETs, where a disturbance might have harsh consequences, it is vital to defend against any such behaviour. Misbehaving nodes can be made to deny forwarding messages that it receives from another member, purposefully misinterpret or modify messages, misuse the available bandwidth or inject bogus information. The technique that has been proposed in [30] can be used to defend against misbehaviour in V2I and V2V communications. It considers the authenticity of anonymous communications in an effort to prevent potential misbehaviour while keeping the privacy of its members. The threshold authentication technique is used where a threshold value is set up to identify strange, unusual or misbehaving nodes a predetermined number of times. If there are any repeated offences, result over the threshold value, it will trace the misbehaving node's credential.

IV. DEFENCES

Existing security and privacy schemes for VANETs can be overall classified into four broad approaches. They all contain different schemes and protocols to ensure the safety of VANETs. The presented protocols main objective is protecting the messages rather than safeguarding the systems themselves against an attacker once it has infiltrated or falsified the transmitted broadcasts.

A. Public Key

A node in the VANET is given a pair of keys, public and secret. A proper Public Key Infrastructure (PKI) should, in theory, be able to efficiently handle the management of keys to provide security. Usually, a scheme that contains the use of a PKI is generally proposed whenever a vehicle has two specific hardware units: Tamper Proof Hardware (TPH) and EDR to perform cryptographic processes and record all the events.

Hesham et al. [33] propose a protocol using a dynamic method of key distribution that can handle key management which does not have a need to store a large number of keys in its memory for employing PKI, thus reducing the amount of TPH needed. Using this approach as a basis, member-specific unique information such as Electronic License Plate (ELP) or chassis number that forms the Vehicle Authentication Code (VAC) can be employed as a hidden key shared by a certificate authority (CA) and a member. CAs are responsible for issuing, renewing, revoking and distributing public keys [34]. The use of this protocol grants a strong resistance to DoS, Sybil and Man-in-the-Middle attacks since it makes use of ELP and a unique VAC encrypted secret key to protect against masquerade attempts.

Gazdar et al. [35] propose an effective dynamic cluster-based architecture for a PKI defence model in VANETs employing a trust model where a value ranging from 0 to 1 is attributed to them. Based on this value, members can have four different roles being CA, Member Node (MN), Registration Authority (RA) and Gateway (GW). RA and CA which are MN that have a trust value that is equal to 1 can issue certificates to the members in the cluster while protecting the CA against an attacker by avoiding any potential direct contact and exchange between the CA and an untrusted vehicle. A GW is employed for inter-cluster communications. Nodes, including MNs, have to demonstrate good cooperation and comportment to increase their values of trust. In this architecture, a hierarchical monitoring scheme is employed to observe the comportment of its members, where a member with higher trust value monitors the vehicles with lower trust values. The PKI based scheme of Efficient Certificate Management Scheme (ECMV) [36], provides an effective certificate management control between different authorities, grating the OBU with the capability of updating its certificates at any instant regardless of its location. Even if an adversary manages to infiltrate the network, an ECMV has a solid certificate revocation protocol to follow to remove the intruder. This scheme has proven to reduce the complexity of managing certificates to a large extent and can be very effective in making PKIs scalable and more secure, which in turn makes it ideal for IoT and VANETs in general.

B. Hybrid and Symmetric

In these type of schemes, nodes start to communicate only after they agree to share a secret key that will be employed for securing communications. Most of the current security schemes available to VANETs are dependent on either symmetric or public keys to encrypt communications. A potential hybrid system that employs both public and symmetric keys has been proposed for securing VANETs [14], using the two different types of communications, being group and pair-wise. During the group communication, more than two nodes can communicate whereas the pair-wise communication is employed when two members need to communicate with one another. The hybrid approach employs symmetric keys for pair-wise transmissions in an effort to avoid the potential overhead that happens when using the key pair. It should be noted that symmetric keys must not be employed during an authentication process as it will prevent non-repudiation which is a main constraint of VANET messages [15].

C. Certificate Revocation

In the broader scope of PKI methodology used to provide security to VANETs, there exists a major category called certificate revocation

[34]. Certificate revocation is usually applied and enforced by the CA in two ways: centralized and decentralized. The first option employs a centralized authority which is only responsible for taking the revocation decisions whereas, in the second option, a group of neighboring members of the vehicle to be revoked take the decision. This scheme is more often than not centralized and requires pervasive infrastructure and thus cannot be said to efficient since RSUs need to send the certificate revocation list to OBUs while the deployment cost is relatively high compared to most other PKI methods. This RSU dependence could be alleviated in the near future with the use of 5G technologies and cloud computing and sharing. For now, a couple of alternative approaches and implementation have been put forward such as the Distributed Revocation Protocol, the Revocation Protocol of Tamper Proof device, the Revocation Protocol using Compressed Revocation Lists. Another newly proposed scheme, the RSU aided Certificate Revocation, where a trusted third party grants different secret keys for each individual RSU so that they can additionally sign the messages transmitted in its range. Once a member's certificate is proven to be invalid, CAs broadcasts messages to the RSUs which in turn transmits orders to all vehicles in range to revoke that particular vehicles certificate and to stop communication with it, effectively isolating the problematic member from the network.

D. ID-Based Cyrptography

PKI and symmetric key cryptography are not always the most optimal schemes to provide security and robustness to VANETs since they are most of the time infrastructure dependent and lacking additional layers of protection. An alternative, the ID-based cryptography, that contains the best features of the traditionally employed security schemes and protocols are being explored by the scientific community. ID-based cryptography reduces the required computational resource cost in the ID-based Signature (IBS) processes and is preferable for using the ID-based Online/Offline Signature (IBOOS) scheme for authentication in VANETs. IBOOS has been shown to increase efficiency by dividing the signing process into two phases, online and offline, in which the verification has been shown to be more efficient than that of IBS.

Lu et al. [37] propose an ID-based framework that makes use of both IBS and IBOOS for authentification. It should employ self-defined pseudonyms instead of any kind of real-world IDs to protect the member's privacy and confidentiality. It has proven to be efficient in term of communication overhead, storage and processing time. This is achieved by preloading a pool of IDs for the overall regional RSUs in each vehicle beforehand, which is often very small in size and is not expected to change as frequently compared to some of the other approaches that make use of pre-stored IDs of all the potential RSUs. This is done by using IBS for V2R authentications while IBOOS is employed for V2V authentications. Evaluation results have shown that this scheme manage to efficiently preserve the privacy of VANETs.

Pan et al. [38] have put forward a model to quantify the locational privacy by applying a simple scheme called Random Changing Pseudonyms. Each member changes its attributed pseudonym after an arbitrary point in time and space. It should be noted that it is very important to provide unlinkability for the two successive pseud-onyms of any member. If this constraint is not fulfilled an intruder might potentially be able to locate the tracked member by mapping the link between the changing pseudonyms. The probability of the pseudonyms unlinkability is directly affected by the efficiency of the various pseudonym shifting schemes in use to protect locational privacy.

V. OPEN RESEARCH ISSUES

The quick development of VANETs has created a great number of open research possibilities, starting with securing them against inside and outside threats. Any research aiming to improve the robustness and reliability of defensive schemes is always open to more in-depth work. The development of smart key attribution be it governmental or local, not linked to IP or MAC addresses, is a necessity as the ever-expanding scalability of the network will most likely not support current identification schemes. The ethi-cal issues should also be discussed in depth and proper liability thresholds should be established to encourage the dissemination and market penetration of VANETs. The matter of fact use of the internet and remote transaction in our daily lives, such as paying a highway toll wirelessly, has made it a concern when dealing with identity theft. This makes the development and design of secured communication protocols for VANETs with an emphasis on protecting user profiles, private data and information from malicious vehicles and infrastructure has to be given a higher priority than before where it was relegated as a secondary concern. A more general topic would be the integration of 5G communication technologies in VANETs and the potential for cloud computing and virtual infrastructure for certificate revocation defense protocols. Another pertinent avenue of research is the scalability of the proposed methods as a defence method which is not designed and stress tested carefully for a high-density and high-traffic situation will most likely fail due to not being able to fulfil constraint or crumble due to errors. Comparison between the efficiency of the four main defence scheme could yield interesting results concerning optimization of VANET security protocols. The analytical models to accurately quantify the effectiveness of pseudonym changing schemes to reinforce unlikability is a major research topic to provide security in VANETs using ID-Based Cryptography as the efficiency of the scheme can depend mainly on this property. Finally, a potential approach that defines trust taking into account the expected behaviour of that vehicle compared to its previous broadcasts and standard movement patterns might be a future defensive scheme.

VI. CONCLUDING REMARKS

The propagation of VANETs will most likely continue and it will become a seamless part of our lives in the future. For this to happen, the major concerns concerning security being privacy, for the users, and robustness, for the health of the network, should be guaranteed and the constant development of new protocols and schemes is a good indicator of how much the scientific community is aware of this reality. The topics of research concerning software development and technical subjects abound as demonstrated in the relevant passage and it should be noted that a lot of such concerns also exist when we take a look at the more physical part of the equation concerning OBUs and how to effectively make them tamper proof. The question of liability and government control and management is a whole other subject that requires the involvement of policymakers and the consensus of the public to answer and establish where responsibility begins and ends. With the advent of potentially self-driving vehicles, establishing an infrastructure, both physical and electronic, where both human controlled and automated members interact harmoniously will become an important factor for the proper implementation of automated vehicles and VANETs.

REFERENCES

[1] Al-Turjman, F., and Alturjman, S. "Context-sensitive Access in Industrial Internet of Things (IIoT) Healthcare Applications", *IEEE Transactions on Industrial Informatics*, 14.6(2018): 2736–2744.

[2] Al-Turjman, F., and Alturjman, S. "Confidential Smart-Sensing Framework in the IoT Era", *The Springer Journal of Supercomputing*, 74.10(2018): 5187–5198.

[3] Andrews, Jeffrey G., et al. "What will 5G be?." *IEEE Journal on selected areas in communications* 32.6(2014): 1065–1082.

[4] Zeadally, Sherali, et al. "Vehicular ad hoc networks (VANETS): status, results, and challenges." Telecommunication Systems 50.4(2012): 217–241.

[5] Alam, Kazi Masudul, Mukesh Saini, and Abdulmotaleb El Saddik. "Toward social internet of vehicles: Concept, architecture, and applications." IEEE Access 3(2015): 343–357.

[6] Sun, Wei. "Internet of vehicles." Advances in Media Technology (2013): 47.

[7] Fangchun, Yang, et al. "An overview of internet of vehicles." China Communications 11.10(2014): 1–15.

[8] Douceur, John R. "The sybil attack." International Workshop on Peer-to-Peer Systems. Springer, Berlin, Heidelberg, 2002.

[9] Guette, Gilles, and Bertrand Ducourthial. "On the Sybil attack detection in VANET." Mobile Adhoc and Sensor Systems, 2007. MASS 2007. IEEE International Conference on. IEEE, 2007.

[10] Raya, Maxim, and Jean-Pierre Hubaux. "Securing vehicular ad hoc networks." Journal of computer security 15.1(2007): 39–68.

[11] Lo, Nai-Wei, and Hsiao-Chien Tsai. "Illusion attack on vanet applications-a message plausibility problem." Globecom Workshops, 2007 IEEE. IEEE, 2007.

[12] Newsome, James, et al. "The sybil attack in sensor networks: analysis & defenses." Proceedings of the 3rd international symposium on Information processing in sensor networks. ACM, 2004.

[13] Sumra, Irshad Ahmed, Iftikhar Ahmad, and Halabi Hasbullah. "Behavior of attacker and some new possible attacks in Vehicular Ad hoc Network (VANET)." Ultra-Modern Telecommunications and Control Systems and Workshops (ICUMT), 2011 3rd International Congress on. IEEE, 2011.

[14] Al-Kahtani, Mohammed Saeed. "Survey on security attacks in Vehicular Ad hoc Networks (VANETs)." Signal Processing and Communication Systems (ICSPCS), 2012 6th International Conference on. IEEE, 2012.

[15] Al Hasan, Ahmed Shoeb, Md Shohrab Hossain, and Mohammed Atiquzzaman. "Security threats in vehicular ad hoc networks." *Advances in Computing, Communications and Informatics (ICACCI), 2016 International Conference on*. IEEE, 2016.

[16] La Vinh, Hoa, and Ana Rosa Cavalli. "Security attacks and solutions in vehicular ad hoc networks: a survey." *International journal on AdHoc networking systems (IJANS)* 4.2(2014): 1–20.

[17] Agah, Afrand, and Sajal K. Das. "Preventing DoS attacks in wireless sensor networks: A repeated game theory approach." *IJ Network Security* 5.2(2007): 145–153.

[18] Tagra, Deepak, Musfiq Rahman, and Srinivas Sampalli. "Technique for preventing DoS attacks on RFID systems." *Software, Telecommunications and Computer Networks (SoftCOM), 2010 International Conference on*. IEEE, 2010.

[19] Mohi, Maryam, Ali Movaghar, and Pooya Moradian Zadeh. "A Bayesian game approach for preventing DoS attacks in wireless sensor networks." *Communications and Mobile Computing, 2009. CMC'09*. WRI International Conference on. Vol. 3. IEEE, 2009.

[20] Verma, Karan, Halabi Hasbullah, and Ashok Kumar. "Prevention of DoS attacks in VANET." *Wireless personal communications* 73.1(2013): 95–126.

[21] RoselinMary, S., M. Maheshwari, and M. Thamaraiselvan. "Early detection of DOS attacks in VANET using Attacked Packet Detection Algorithm (APDA)." *Information Communication and Embedded Systems (ICICES), 2013 International Conference on*. IEEE, 2013.

[22] Tippenhauer, Nils Ole, et al. "On the requirements for successful GPS spoofing attacks." *Proceedings of the 18th ACM conference on Computer and communications security*. ACM, 2011.

[23] Hu, Y-C., Adrian Perrig, and David B. Johnson. "Packet leashes: a defense against wormhole attacks in wireless networks." *INFOCOM 2003. Twenty-Second Annual Joint Conference of the IEEE Computer and Communications. IEEE Societies*. Vol. 3. IEEE, 2003.

[24] Safi, Seyed Mohammad, Ali Movaghar, and Misagh Mohammadizadeh. "A novel approach for avoiding wormhole attacks in VANET." *Computer Science and Engineering, 2009. WCSE'09. Second International Workshop on*. Vol. 2. IEEE, 2009.

[25] Boneh, Dan, and Matt Franklin. "Identity-based encryption from the Weil pairing." Advances in Cryptology—CRYPTO 2001. Springer Berlin/Heidelberg, 2001.

[26] Park, Soyoung, et al. "Defense against sybil attack in vehicular ad hoc network based on roadside unit support." Military Communications Conference, 2009. MILCOM 2009. IEEE. IEEE, 2009.

[27] Raya, Maxim, Panos Papadimitratos, and Jean-Pierre Hubaux. "Securing vehicular communications." IEEE Wireless Communications 13.5 (2006).

[28] Xiao, Bin, Bo Yu, and Chuanshan Gao. "Detection and localization of sybil nodes in VANETs." Proceedings of the 2006 workshop on Dependability issues in wireless ad hoc networks and sensor networks. ACM, 2006.

[29] Sun, Jinyuan, and Yuguang Fang. "A defense technique against misbehavior in VANETs based on threshold authentication." Military Communications Conference, IEEE. MILCOM 2008.

[30] Alabady, S., Al-Turjman, F., and Din, S. "A Novel Security Model for Cooperative Virtual Networks in the IoT Era", *Springer International Journal of Parallel Programming*, 2018. DOI: 10.1007/s10766-018-0580-z.

[31] Chim, Tat Wing, et al. "Security and privacy issues for inter-vehicle communications in VANETs." Sensor, Mesh and Ad Hoc Communications and Networks Workshops, 2009. SECON Workshops' 09. 6th Annual IEEE Communications Society Conference on. IEEE, 2009.

[32] Chim, Tat Wing, et al. "SPECS: Secure and privacy enhancing communications schemes for VANETs." Ad Hoc Networks 9.2(2011): 189–203.

[33] Hesham, Ahmed, Ayman Abdel-Hamid, and Mohamad Abou El-Nasr. "A dynamic key distribution protocol for PKI-based VANETs." *Wireless Days (WD), 2011 IFIP*. IEEE, 2011.

[34] Al Falasi, Hind, and Ezedin Barka. "Revocation in VANETs: A survey." *Innovations in Information Technology (IIT), 2011 International Conference on*. IEEE, 2011.

[35] Gazdar, Tahani, Abderrahim Benslimane, and Abdelfettah Belghith. "Secure clustering scheme based keys management in VANETs." *Vehicular Technology Conference (VTC Spring), 2011 IEEE 73rd*. IEEE, 2011.

[36] Wasef, Albert, Yixin Jiang, and Xuemin Shen. "ECMV: efficient certificate management scheme for vehicular networks." *Global Telecommunications Conference, 2008. IEEE GLOBECOM 2008. IEEE*. IEEE, 2008.

[37] Lu, Huang, Jie Li, and Mohsen Guizani. "A novel ID-based authentication framework with adaptive privacy preservation for VANETs." *Computing, Communications and Applications Conference (ComComAp), 2012*. IEEE, 2012.

[38] Pan, Yuanyuan, et al. "An analytical model for random changing pseudonyms scheme in VANETs." *Network Computing and Information Security (NCIS), 2011 International Conference on*. Vol. 2. IEEE, 2011.

4

SEAMLESS IDENTIFICATION IN IoT

FADI AL-TURJMAN AND CHADI ALTURJMAN

Contents

Abstract

The Internet of Things (IoT) represents an evolutionary vision and the new era of such smart environments that encompass all identifiable things in a dynamic and interacting network of networks. As each user has wide interaction with a huge number of entities. It would be impractical to require people to verify themselves every time they cross various network boundaries, as the frequent verification process would disrupt the users' normal activities and degrades the overall performance. This paper presents a Seamless Identification Method (SIM) provisioning framework for IoT that relies on RSNs technology for user identity verification in a non-intrusive method through proposing the idea of monitoring and inferring certain types of user activities in smart environments. This framework helps in supporting the creativity of the IoT by being smart, boundless, easier and safer era to improve people lives. The proposed framework reduces the risk of identity theft which results from losing user devices that include an RFID tag, where the user identity is usually stored. It will cope with the loss of the user's ID or people impersonating other people and raise an alarm to block an intruder from being verified as a legitimate user.

Keywords

Internet of Things (IoTs), Information-centric networks, Smart environments, RFIDs; Sensor networks; User identifiction, Authentication.

I. Introduction

The Internet of Things (IoT), in a nutshell, is a network of everyday objects. Identifying and networking all the objects in the world establish a new paradigm of digital interactions with the physical world established; presenting a revolutionary transformation in our daily lives [1]. Radio frequency Identification (RFID) and wireless sensor networks (WSNs) are the two key enabling technologies in the creation of IoT. With the recent advancement of smart computing environments, Internet of Things (IoT) paradigm has become an integral part of many aspects of our daily lives from connected homes and cities to connected cars and roads to devices that could follow individual's behavior [2]. However, people have limited time, attention and accuracy that mean they are not so conscious and good at capturing data about things in the real world. Therefore, the emergence of computers and smart embedded sensors networks (SN) that knew everything about surrounding things, using data they gathered without any help from people, would be able to track and count everything and greatly solve many of such capabilities like decrease providing ubiquitous access of various data to people at lower costs and reducing wasting time. As this promising computing era is a hot topic and developing fast, many research opportunities have just started to unveil. Thus, we shall highlight challenges for seamless and pervasive services and applications provisioning.

In the IoT, connectivity between devices is embedded in an unobtrusive way and always available in everything and everywhere. It changes and improves not only the human work environment but also daily life and communication patterns between families/friends [3]. There are many advantages for implementing such scenarios, such as moving the interaction with computers out of a person's central focus into the user's peripheral attention, where they can be used subconsciously. Another creativity of IoT is to make our lives more comfortable by providing device mobility and digital infrastructure, and the ability to provide useful services for people [4]. Moreover, a user has various connections with many smart devices, regardless of the hardware specifications or the software restrictions.

As users dealing with various integrated smart entities such as, Radio Frequency Identification readers (RFID) and Wireless Sensor Networks (WSN), and software applications in addition to other people as well, a model to address the intrusive problem interaction is required. Such a model would help users to avoid and control intrusive interactions, reduce users' disturbance, and allow them focus on their tasks. However, such model introduces various risks and security issues especially with user identity and privacy. Therefore, and in order to provide more flexibility in offering a variety of services through IoT, it is essential to ensure and protect the user identity [5].

Normally, people establish their identities using traditional approaches, such as biometric scanning, or by camera, that are used to track behavior. However, we propose an idea of non-intrusive monitoring of certain user activities from which we can assert his/her identity. This method is not extensive behavior tracking, but it is only monitoring selective user login activities. Thus, the idea of non-intrusive authentication procedures will offer enhanced user experience and improved provision of web-services.

In this paper, we propose a Seamless Identification Method (SIM) identification method that certifies user identity to provide him/her seamless transfer from one web-service to another while connected to the Internet.

II. Related work

Implementing user access control in IoT requires innovative techniques to establish the identity of users who wish to access various web-based resources within such typical smart environments. Users in such environments use various mobile devices and technologies such as RFID tags, smart tags or smart phones when interacting with the environment. Their identity is stored in the device's memory and would be retrieved and used to identify users every time a user attempts to access a resource or service. Some of our related research activities have been in the area of distinguished various approaches in future networks including Internet of Things (IoT) [7][8][24]. The approach of using sensor nodes and other enabling technologies as service providers in the IoT era has seen some progress. In this approach, a node would be connected to a network, which is connected directly/indirectly (via other network) to the Internet, either in a single hop or through relays. Accordingly, all the sensed data would be sent to a centralized database for later processing. However, these systems suffer from identity verification and validation issues dictated by the different kind of networks and enabling technologies and in many cases they depend on higher end nodes with significant power and processing capacity to facilitate authentication to the end-user.

It is important for IoT paradigm to be able to verify the identity of users and at the same time maintaining minimum interruption (caused by the verification process) to their activities. Existing identity verification approaches have focused mainly on four methods; these are (1) biometric technologies, (2) using a certification authority, (3) password exchange protocols based on encryption and (4) context aware identification. In the first method using biometric technologies such as retinal scan, face recognition or fingerprint, pattern recognition techniques are applied to measurable physiological or behavioural characteristics [6]. Such systems verify the user's identity using the biometric characteristics to accept (or decline) the user's claim to their identity [9]. Although such systems offer features such as accurate identification, universality, permanence and uniqueness [10], they pose considerable challenges to IoT world systems. Such challenges are privacy intrusion; the need for additional hardware and software to acquire the information; and user acceptance and perception. In the second method using certification authority, a number of solutions have been proposed for this purpose. Some employ the use of a proxy server in order to give users access to secure proxy credentials when and where needed, without requiring them to directly manage their long-lived credentials [11]. Others use dedicated identity service such as cloud services to provide the identification method required as a verification agent [12]. The use of services, proxies or agents provide the convenience of centralised access control of resources, but place an extra infrastructural requirement on the whole system design. It also requires effective solutions to intrusion detection and Man-in-the-Middle (MiM) problems.

There are currently a number of solutions that rely on the use of password exchange protocols (method 3), which are based on the use of the SIM concept [13][14]. Some use standard technologies such as Kerberos, OpenID or Security Assertion Markup Language SAML [15][16][17], while others define proprietary protocols such as the Identity Service Provider Protocol (ISPP) [18]. These approaches have the ability to use identity information from one domain to access resources in another domain based on trust relationships that have to be established previously [19]. This approach however has scalability issues due to the fact that complex trust relationships may soon evolve into a complex map of inter-domain relationships.

In IoT scenarios, identity verification takes a step further by utilising the system's knowledge of its users; this is commonly known as context information [20]. Context-aware identification systems (method 4) use knowledge of the current status of the user to verify the user's identity [21][22]. Such systems utilise context information that is relevant to the user and hence has the potential to be scalable and less dependent on complex trust relationships. However, there are a number of issues with this approach. Firstly according to [20], identity is one of the main pillars of context information and to use context to verify it seems cyclical. Secondly, context information is dynamic in nature and hence using it as a basis for access control may cause serious issues and require complex policies for allocating and revoking group memberships [22]. Thirdly, users are usually less inclined to surrender personal information related to their current status in the process of verifying their identity.

In this paper, the presented framework provides a user-centric approach that supports scalability; each individual identity is treated in isolation of others regardless of the domain to which the resource belongs. In other words, the framework isolates the process of verification from other authentication and authorisation procedures, which makes the scalability issue easier to manage. In addition, this framework offers minimum interruption to the user's activities. The framework, termed Seamless Identification Method (SIM), uses history of the user's previous access to the environment's resources as a means for asserting the user's identity before access control rules are applied.

The rest of the paper is organized as follow: next section, the proposed approach architecture and notations are described then discussing its principles, components and operation. In section 4, the approach used in identifying and classifying user activities is explained. This is then followed by a section discussing the system's testing and tuning; this section discusses how the system has been tested and tuned for various security settings. Then, the evaluation and validation of the system is discussed based on the use of discrete event simulation which was developed specifically to evaluate the behaviour of the system. Finally, a number of conclusions are discussed highlighting the achievement and limitation of this work.

III. System Architecture and Notations

The proposed system certifies user identity to provide him/her seamless transfer from one service to another while connected to the crowd of sensors and RFID systems. In this section we describe the SIM notations and the architecture.

A. Notations

In the SIM implementation, testing and tuning a number of parameters used in asserting the user identity in a non-intrusive approach. Table 4.1 shows the clarification of each parameter and equation:

Table 4.1 SIM Notations

Global assertion value	g
Current activity	A
New global assertion value $g_i \in [0,1]$	g_i
Initial global assertion value	g_0
The assertion threshold $g_{tj} \in [0,1]$	g_{tj}
The local assertion value such that: $l \in [-1,1]$	l
The rate of change of the global assertion value	Δg
An integer loop control variable such that: $i \in [1, \infty]$	i
Recognised recurrent activity	A_r
A sequence of recognised activities	A_q
A significant activity	As
Identifying activities	A_{is}
Repudiating activities	A_{ps}
Unrecognised activity	A_u
No activity	A_0
Assigned value for -no events is	$-K$
Assigned value for recurrent events is	K_r
Assigned value for significant events is	K_s
Assigned value for sequence of recurrent events is	K_q
User's id	Id
Resource location	L
Time	t
Euclidean distance measure	d
Minimum Cluster size	z
Time tolerance	ε
Event probability	p_e

B. SIM Architecture

The architecture of the SIM is shown in Figure 4.1. It consists of the following stages; the first stage is the smart sensing (acquisition) stage in which user activities detected through integrating RFID readers with sensor nodes. Here, the existence of three types of devices is assumed to be as follows: the integrated RFID reader-sensor nodes, simple RFID tags, and the sink or base-station. The collected information is about the individual users' positions and movements. The user activities are stored in a user activities database as a history about the user. Then the second stage represents the Activity Mapping and Classification, where user activities are identified as events. Within this component an event is classified into a specific type of events, and added to the user events database. For example, when a user attempts to access a resource by his/her smart card, the system identifies this action as a user activity, which it uses to verify the user's identity. The system maintains two databases for this purpose; one is called the activity database which stores a history of all user activities as they access various resources in the environment, and the second is called event database which stores recognized recurrent activities. The third stage is designing an algorithm, where the algorithm is used by the smart framework to analyse the activity and calculate a numeric value, which reflects how much confidence the framework has about the

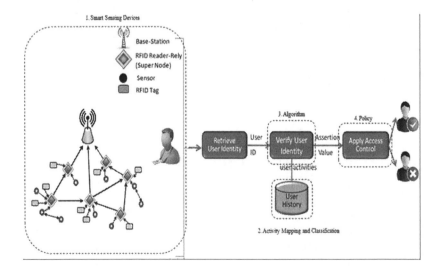

Figure 4.1   SIM Architecture.

user identity. The fourth stage includes the policy that is used by the smart framework to make a decision whether to grant the user access to a specific resource or not, based on a calculated system confidence numeric value. The system will then use this value to decide whether the user can be trusted to access the resource or not based on a policy.

The proposed system uses the following three principles to verify the user identity:

1. Identification: the user will first be identified using a personal identification device such as a mobile device or a smart tag (RFID tag) which is scanned by a suitable reader attached to the protected resource.
2. User activities: The system uses patterns of previous access to resources (history) in order to assert their identity. For example, is the user known to have been accessing a certain resource at a particular time?
3. Asserting user identity: The verification of user identity is stepwise and leads to levels of assertion (confidence) that is represented as a numeric value. The numeric value signifies whether the system is confident to allow the user access to that resource.

In SIM, identity verification is a process of monitoring users as they attempt to access the environment's resources and convert this knowledge into a numeric value (assertion value) that represents the system's confidence of the user's identity. This value is computed using an algorithm called the assertion algorithm that is used to update the system's confidence as the user performs a known activity or otherwise. The amount by which the system's confidence is altered depends on the type of activity that the user performs. SIM classifies activities into a number of types, and uses these types in calculating the assertion value instead of the actual activities. This reduces the need to record details of the user activities usually used in the interpretation process. The system will then use the assertion value to decide whether the user can be trusted or not based on a policy known as the assertion policy.

The SIM operates on the basis of assertion levels. If the system confidence about a user reaches the assertion level needed to gain access to a specific resource, then the user identity is said to be verified.

In order to achieve this, it uses an algorithm that applies the following principles:

1. For each user, it maintains a numeric value known as the global assertion value (g) which reflects how confident the system is about the user's identity at any time.
2. It divides the protected resources into a number of assertion levels (n). In each level, the system keeps a value called the assertion level threshold (g_{t1}, g_{t2},..., g_{tn}) that reflects how confident the system must be about a user in order to assert their identity before they are granted access to the resource.

The global assertion value (g) is updated continuously using Equation (4.1).

$$g = f(g', A) \tag{4.1}$$

The current activity is represented using a numeric value called the local assertion value (ℓ) which represents how much a particular activity can influence the system's confidence about the identity of the person performing the activity. It determines the direction and extent by which (g) swings as a result of that activity. The new global assertion value is calculated using Equation (4.2).

$$g_i = \ell * \Delta g + g_{i-1} \tag{4.2}$$

The local assertion value is calculated for each activity performed by the user. The algorithm also defines a control parameter representing the rate of change of the assertion value and referred to as Δg. This parameter is used to control the speed by which the assertion value increases or decreases. In addition, the SIM classifies the user's activities into five types for the purpose of calculating the local assertion value. Below are the classified activities:

1. A recognised recurrent activity (A_r): This type of activity is an activity that occurs repeatedly (at a specific time). These activities have a moderate influence on the system's confidence.

2. A sequence of recognised activities (A_q): This type involves sequences of activities which occur periodically and in succession. Activities of this type have more influence on the system confidence than the first type.

3. A significant activity (A_s): This type of activity is related to specific activities users normally perform and involves some kind of positive identification. These activities are important to the system because they can sharply raise or lower the system confidence. They can be further classified into two sub categories:

   • Identifying activities (A_{is})
   • Repudiating activities (A_{ps})

4. Unrecognised activity (Au): This type is when the user performs an activity that the system does not find in the activity database.

5. No activity (A_0): The system does not detect any user activity over a period of time.

Each type of activity has different effect (weight) on the system's assertion value. For example, when a user performs a recognised activity such as accessing a particular resource at a specific time, the system's confidence will elevate to a higher level of assertion.

Furthermore, the assertion policy comprises setting a number of parameters that determine the behaviour of the SIM verification process, and consequently affects the decision when to grant the user access to the protected resources. The following parameters are specified in the policy:

1. Assign the initial global assertion value, this could either be 0, 1 or any other value depending on the environment's security policy.

2. Determine the local assertion value for the various activity types. As SIM identifies one of the activities, it calculates the local assertion value as a function of the activity type, and as specified in the assertion policy. The policy defines a number of parameters as shown in Equation (4.3). The parameters are typically assigned by an administrator depending on how strict the security policy is required in the environment.

$$\ell = f(A) = \begin{cases} 0, & A \in A_u \\ -K, & A \in A_0 \\ K_s, & A \in A_{is} \\ -K_s, & A \in A_{ps} \\ K_r, & A \in A_r \\ K_q, & A \in A_q \end{cases} \tag{4.3}$$

3. Assign the value of Δg, as an absolute value, which determines the speed of convergence of the whole verification process. This parameter determines how quickly the assertion value decays to zero if no activity has been detected.

4. Identify the number of assertion levels (n) needed for the environment, and the assertion threshold for each level. Divide the resources in the environment between these levels such that higher the security resource, the higher the assertion level, and vice versa. The threshold values may be assigned by the administrator to any values; however, SIM recommends the method shown in Equation (4.4) for calculating the threshold values which is a function of (n).

$$g_{tj} = (j-1) * \frac{1}{n} \tag{4.4}$$

The assertion policy involves tuning the values of a set of parameters as defined in the assertion policy vector matrix, Equation (4.5).

$$Assertion\ policy = [n,\ g_{ti},\ \Delta g,\ g_o,\ K,\ K_r,\ K_q,\ K_s] \tag{4.5}$$

$$Such\ that : j \in [\,1,\,n\,],\ K_r \le K_q \le K_s\ and\ g_{t1} \le g_{t2} \dots \le g_{tn}$$

Although, the assertion policy allows the administrator to choose any values for these parameters to suit their specific security requirements, SIM provides a number of pre-determined sets for aggressive, relaxed and moderate securities.

IV. Seamless Identification Method (SIM)

This section details the design and development of the Cloud-based system through a pre-deployment phase and three operational phases. The pre-deployment is an experimental phase in which we train the system [15][16]. Operational phases include: i) Client-side service request by smart phones/PCs which have Internet-based user accounts, ii) Virtual profile creation using the aggregated data on the cloud, and iii) User verification/authentication via virtual ID.

A. Pre-Deployment Training

This phase is experimental in nature, involving the collection of user-related data from a number of smartphones/PCs that are located inside the vicinity of the user. The data collected from the smartphone/PCs is transmitted to an analytics platform for storage, pre-processing, and analysis purposes [17][18]. Training algorithms are also applied to the collected data. As part of the system training, we developed a unique analytics suite on a secure Virtual Profile Platform by using cloud service for Virtual Profile Creation.

Our system is expected to work via the orchestration of application and internet service providers' component that is installed by the users wishing to participate in this process together with an analysis and visualization platform that can be deployed by a third party (e.g., government offices or any other interested party). The bulk of analysis and aggregation is performed at a data analytics platform with two deployment choices: either private in-house deployment by the third party, or cloud deployment that can be public or private, depending on the degree of privacy desired for the outcomes. In either case, the third party can choose to make the results of the analysis platform available to third parties for potential collaboration and investment opportunities. The anticipated cloud-based system architecture is illustrated in Figure 4.2 and is briefly explained as follows:

B. Data Acquisition/Client-Side Service Request

The functionalities that take place in this part of the system are carried out at the smart phones/PCs which have Internet-based user accounts

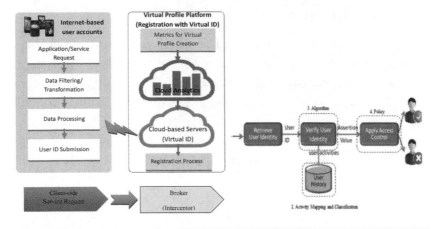

Figure 4.2　Cloud-based architecture for the virtual profile creation.

and are concerned with application/service request, data processing, and user ID submission by using client-side service request to the Virtual Profile Platform. The Application/Service Request module is needed to collect and combine data generated by the accessed applications. The data filtering/transformation module applies smoothing algorithms to the collected data in order to remove jitter, and transforms the data into a usable format suitable for further processing/analysis. The module also handles the temporary buffering of data as it is being processed by the data processing module. The data processing module identifies the data instances corresponding to the users/events of interest to the system. The classifiers produced and fine-tuned in the training phase are used in the data processing module to perform this identification and discard "white noise"; readings that most probably do not reflect events of interest to the proposed system. The user ID submission module sends a report of events to Cloud-based servers associated with the third party. A design decision is made at this point with regard to the time at which data is reported (i.e. immediately or in batches). The technology to be used for data transfer is chosen according to a weighted utility function that dictates the choice of medium – Cellular (LTE, 4G/3G), Wi-Fi, or SMS – used for transmitting data from smartphones/sensing-cloud to the analytics platform. At the end of this module, registration process occurs.

C. Virtual Profile Platform/Data Analysis

This part is where data-intensive analysis techniques are deployed in order to continuously aggregate and provide insight and validation into the reported events. The core of this phase is the data analytics platform, which is realized via a suite of algorithms that are developed and offloaded to a cloud solution. Data that is reported by PCs/smartphones is stored in a Cloud-based server that can be accessed by the data analytics component in the platform.

Data analytics in the operational phase have a slightly different task than its counterpart's in the training phase. As in the analysis phase, analytics in the operational phase involves the deployment of cloud-hosted in-house analytics services in which parts of the data analytics process (e.g. data models, processing application, analytic models, and storage of results) are provided through a public or private cloud. However, the role of analytics in the operational phase is to provide real-time and updatable aggregation, filtering, tagging, and prioritization of the events reported by the users participating in the proposed system. Of special importance to the analytics process is information tagging, which involves assigning a priority and lifetime to the event based on its type and criticality, and information geo-referencing, which involves associating each event/access with its geographical location, as well as identifying a time at which the event took place. Events geo-referencing is performed with the aid of GPS, cellular towers and/or Internet service providers.

After the events are tagged and referenced, they are evaluated using the Quality of Information metrics. The analytics platform handles the scalability of data that can be used for further analysis, via mechanisms such as sampling or sliding window aggregates. Further analysis is needed in cases where the detected access/events that are sent by users do not have enough support, indicating the event presence of novel false triggers that were not caught in the training/analysis phase. Constant fine tuning of the system to filter out novel patterns of false triggers – that are detected but do not reflect events of interest to the monitoring system – is needed in order to maintain a high accuracy rate for the system's detection mechanism. In addition, incorporating social reporting into analysis can help with the assessment of the features/events severity.

By using the proposed approach, we were able to securely and seamlessly authenticate users and allow them to use distinct web applications. A schematic diagram of the SIM/Seamless Identification Method (SIM) system architecture illustrating its operation is shown in Figure 4.3. When a user attempts to access a resource (for example by scanning their smart card), the system identifies this action as a user activity, which it uses to verify the user's identity. The system maintains two databases for this purpose; one is called the activity database which stores a history of all user activities as they access various resources in the environment, and the second is called event database which stores recognised recurrent activities. Activities are represented as a vector of user identity, resource location and time of accessing the resource as described in Equation (4.6).

$$A = [id, L, t] \tag{4.6}$$

In the activity classifier and mapping part, the system applies an unsupervised clustering algorithm [23] to identify activities as events and maps them into one of the activity types. If the system recognises the activity as a recurrent, significant, or sequence of activities, it adds it to the events database. Here, we use the term event to denote a recognised user activity. If the system identifies the detected activity as an event that already exists in the events database, it determines its type and passes it to the controller.

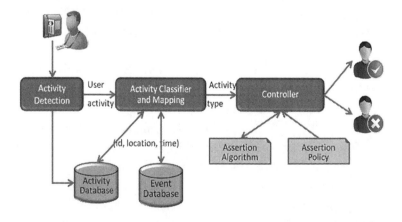

Figure 4.3   SIM framework.

The controller uses the type of event to calculate the local assertion value according to the assertion policy. It then applies the assertion algorithm (Equation [4.2]) in order to compute the updated global assertion value. Based on this value, the assertion level of the resource and the parameters specified in the assertion policy (Equation [4.5]), the controller determines whether the user's identity is verified to access the resource or not.

D. Identification of Event Types

As discussed in the previous section, central to the operation of SIM is identifying activities as history of recognised events and mapping them into types. To convert activities into events in order to compute the local assertion value (Equation [4.3]), SIM uses a clustering algorithm that works by grouping related user activities to form clusters, which may be identified as events. Due to the dynamic and unpredictable nature of user behaviour, an unsupervised clustering technique is used. Since users do not behave in a uniform manner, the clustering technique is expected to produce an unidentified number of clusters (events) for each user. Therefore, the unsupervised approach is a suitable choice as it does not require any pre-assumptions about the data.

The clustering technique used in this approach is based on the nearest neighbour algorithm [23], and uses the Euclidean distance measure between two user activities as a criterion to measure the similarity between two activities. This can be determined by calculating the 2-norm of the difference between the two vectors; Equation (4.7).

$$d = \|A_2 - A_1\|_2$$

$$= \sqrt{(id_2 - id_1)^2 + (L_2 - L_1)^2 + (t_2 - t_1)^2} \qquad (4.7)$$

Assuming both activities are performed by the same user and at the same location in order to be similar, Equation (4.7) can be simplified as shown in Equation (4.8).

$$d = |t_2 - t_1| \qquad (4.8)$$

Two more quantities are used by the clustering algorithm in the process of grouping activities into events:

1. Time tolerance (ε): is the distance within which a group of user activities may be grouped to form a cluster. Time tolerance is used to ascertain that all points in a cluster must be sufficiently close (or similar) to one another.
2. Minimum cluster size (z): is a minimum number of activities (presented as ratio) that a cluster must contain in order for that cluster to be accepted as an event; see Equation (4.9). It is assumed that a group which contains a certain number of activities that is greater than or equal to z, is recognised as an event and added to the events database.

$$z = \frac{\text{minimum number of activities in a cluster}}{\text{number of all possible occurences}} \qquad (4.9)$$

Using ε, the clustering algorithm groups activities into a number of clusters of varying sizes. Depending on the size of a cluster, not all clusters can be recognised as events. To decide whether a cluster represents an event or not, the cluster size (s) has to be determined. This is calculated as an event probability (Pe), as shown in Equation (4.10).

$$s = P_e = \frac{\text{number of activities}}{\text{number of possible occurances}} \qquad (4.10)$$

Such that if $P_e \geq z$, then the activity is added to the events database. Otherwise, the activity is not recognised as an event, but kept in the activity database to give the system the ability to recognise the activity if it is repeated in the future.

Identifying events is the principal method for identifying recurrent and sequence event types. Once events of a specific user have been identified, a Markov chain model is used to measure the likelihood of two or more events taking place in a sequence. Identified sequences of events are added to the events database, with indication of the sequence degree (the number of events in the sequence).

Table 4.2 Estimated Clustering Parameter Values

PARAMETER	DESCRIPTION	ESTIMATED VALUE
ε	Time tolerance	20
z	Minimum cluster size	0.5

This degree is used when this type is mapped by the Mapping function and by the controller when calculating the local assertion value. Significant events are not subjected to the probabilistic methods as they are inherently identified and each significant activity is recognised as an event. Once a significant event has been detected and used in calculating the assertion value, it will be removed from the events database.

For the purpose of implementing the algorithm, the values of ε and z needed to be estimated. We conducted an experiment to study and analyse patterns of user activities as they access various parts of a building, where they worked. We found that users, who repeatedly access the same part at the same time of day, tend to access that resource within a time span of $\pm\,20$ minutes. We also found that users tend to repeat the same activity at least 50% of the time, for that activity to be recognised as a recurrent activity. This is based on a study of cycles of five working days within a week. Therefore our implementation used the values shown in (Table 4.2).

V. Performance Evaluation

In this section, SIM has been tested, tuned and evaluated extensively using a simulated environment. Simulation allows us to conduct tests using a large number of scenarios involving a variety of activity patterns which is impossible to do within a real environment. The simulator is written in C++ and uses a time driven DES strategy where time is represented as an advancing clock. Events are generated by simulating the time interval between event arrivals as a random variable. All random variables used in the simulation are assumed to fit a normal probability distribution, although other distributions such as Poisson and Exponential distributions have been considered. The simulator was used to tune the assertion algorithm parameters $(\Delta g, K_r, K_s, K)$ for a number of trials based on values calculated from the collected real data. The purpose of the tests was to tune the system's parameters for a number of assertion policies. This has been accomplished successfully and a number of settings have been suggested and presented as guidelines for recommended assertion policies. The purpose of the evaluation was to measure the system's reliability for various types of environments. Furthermore, a scenario is presented to describe the SIM and show its requirements. The scenario is considered a good way to show hypotheses in designing a system it illustrates the requirements of the proposed approach and helps to understand how this approach would behave in a particular situation.

A. Use Case

The following overview includes a scenario presented as a case study to reflect the hypothesis of the proposed framework and the system behavior for a user assumed to carry his ID all the time. The principles of the method and components are also explained to identify the capabilities of the proposed approach. One of the main components is the assertion algorithm, which is used for verifying user identity in a non-intrusive way before issuing a decision to allow access or not to a resource. We developed an approach for monitoring certain types of user activities and not all his/her activities. We present a scenario taking place at a University to describe the non-intrusive identity mechanism and show its requirements. This scenario illustrates the requirements of the proposed approach and helps to understand how this approach would behave in a particular situation. To evaluate the work we presented several scenarios to understand the behavior of the algorithm. Below are the scenarios that we proposed.

At some point in the future the School of Computing, Science and Engineering at Guelph University has developed its facilities and installed many invisible and intelligent devices which are located in different places to create a friendly environment. The school does not wish to ask people to enter their passwords over the identification equipment to establish their identities. This environment will present a smart space that has the ability to interact intuitively with a user and offer him/her easier access to services. Therefore, the University has provided all students and staff with smart tags to ascertain their identity in the school building and anywhere else within the University campus.

There are RFID readers and sensor networks which are located in different places in the school (e.g. room entries and communal places). There are areas within the school that require higher security than others and these areas may be anywhere in the school regardless of their geographical location. For example, public places do not need as strict security as restricted places. The decision has been taken to divide the school facilities into three security levels which are not geographically related. These three levels are: low, medium and high security levels. An example of low security level is the main entrance of a building or refectory. Another example of the medium level could be a laboratory or the postgraduate office (CNTR) while the high security level's example could be the mail room. In addition the security levels also reflect the required level of trust in the user identity. This is how confident the system should be about the identity of the person who carries the smart tag, in order to grant/deny an access to that area.

As a user made activities within the smart environment, after a period of time (two months) the system has collected enough information about the user to recognize his identity. Some activities will prove more significant than others, such as the library login, the supervisor meeting and his PC login activities. As a result, the system has the ability to confirm his identity in various places through history related to such user activities as the time of accessing specific places (CNTR office, café shop, and mail room), the login activity, and the supervisor meeting. The system will make a comparison between previous activities and recent activities to confirm user identity. These comparisons will have an effect on the confidence in a user identity as it might be either increased or decreased.

B. Simulation Results

The system has been tested using a simulated environment in order to tune its behaviour and measure its reliability. The study showed that a number of parameters must be carefully chosen and tuned so that the system would implement various security settings for strict, moderate or relaxed security environments. The following situations were studied using various security settings:

1. Determining the time period that the system waits before it assumes that no activity is detected.

2. Studying the effect of changing the parameters specified in the assertion policy; Equation (4.5). Particularly, studying the effect of Δg, K, Kr, Ks parameters.
3. Studying the effect of the distribution of resources over various assertion levels in the environment.

The testing strategy has taken into consideration the system's behaviour with respect to changing the above parameters. By fixing one parameter and changing the values of the others, the tests were repeated several times in order to gain understanding of the system's behaviour and determine the optimum values for various security settings.

All tests used in evaluating the approach assumed the number of assertion levels n=3. The threshold values of the assertion levels are thus calculated according to Equation (4.4), and shown in (Table 4.3).

1. No Activity Time Period Several experiments were conducted in order to study the effect of updating the global assertion value when no activity is detected using periods between 10–120 minutes for different values of Δg. It was found that using short periods (such as 10 minutes) will cause the system to lose confidence too quickly and produce a large number of false positive alarms. Using long periods (such as 90 or 120 minutes) will cause the system to be too slow to update its confidence, and hence fails to detect any fraudulent identity. We found that (50–60 minutes) would produce a reasonable response for the system.

2. The Effect of Δg on the System Response Changing the value of Δg affects how the system confidence is swayed and illustrated by using a series of recurrent event types happening at equal time spaces (Figure 4.4). As shown in the figure when Δg is set to values between $0.1 - 0.5$, the system confidence increases slowly and took a minimum of six events for the global assertion value to reach the next threshold

Table 4.3 Assertion Threshold Test Values

PARAMETER	VALUE
g_{t1}	0.00
g_{t2}	0.33
g_{t3}	0.66

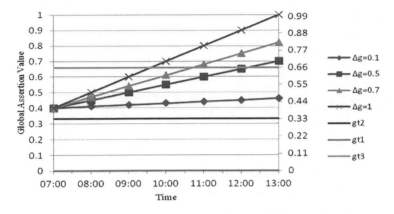

Figure 4.4 System response for recurrent events with different Δg settings (Kr = 0.1).

value. This setting would be useful in highly secure (strict) environments. However, when Δg has high values such 0.7 – 1.0, then the system confidence rises quicker and fewer events are needed to reach the second level (around 2–3 events); this setting would be useful for more relaxed environments.

In (Figures 4.5 and 4.6) the results of testing the identifying and repudiating significant events are shown (respectively). The figures show big differences between the gradients for Δg = 0.1, 0.5 and 1. In the case of the identifying event (Figure 4.5), and for Δg value 0.1, the assertion value rises slowly and takes around three identifying

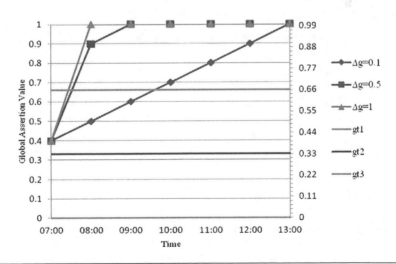

Figure 4.5 System response for identifying events with different Δg settings (Ks = 1.0).

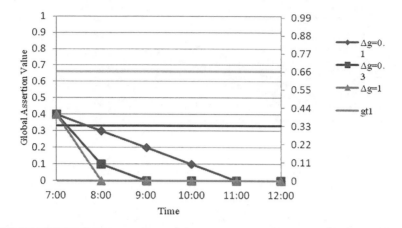

Figure 4.6 System response for repudiating events with different Δg settings (Ks = 1.0).

significant events to reach the next assertion level; this is considered to be unfavourable as it means the system would not trust the identifying events. For $\Delta g = 0.5$ and 1, the assertion value increases more quickly and takes one identifying significant event to achieve the same effect, which is more preferable.

3. The Effect of the Event's Assertion Value on the System Response Choosing different settings to represent the local assertion values for recurrent events will produce similar responses for strict, moderate and relaxed settings (Figure 4.7). As shown in the figure, choosing

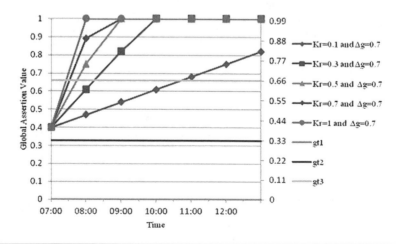

Figure 4.7 System response for different recurrent event settings.

settings in the range of (Kr = 0.7–1.0) will make this type of event to have an effect similar to that of the identifying significant event type. Therefore, Kr is recommended to be assigned a value between 0.1 – 0.5.

SIM has been tested and evaluated extensively using a simulated environment. Simulation allows us to conduct tests using a large number of scenarios involving a variety of activity patterns which is impossible to do within a real environment. The purpose of the tests was to tune the system's parameters for a number of assertion policies. This has been accomplished successfully and a number of settings have been suggested and presented as guidelines for recommended assertion policies. The purpose of the evaluation was to measure the system's reliability for various types of environments. The evaluation showed that the approach is successful in verifying user identity with high degree of reliability. The evaluation results also showed that the approach is particularly effective in certain environments that satisfy the following conditions:

1. The environment has a combination of different security level resources, such as high, medium and low security.
2. The number of low security resources is greater than the number of high security resources in the environment.
3. The user is assumed to perform at least one significant event type per day, such as logging in.

According to the conditions above the system may not perform in the same effective way in such environments where all resources are equal in their security level, or when no significant events can be identified.

VI. Conclusion

In this paper, the presented method (known as SIM) is based on using history of previous access to resources as a means for asserting the user's identity before access control rules are applied. Minimum monitoring of user activities has been used as certain types of activities are identified and recorded, thus minimising the possibility of intrusion on normal user behaviour. In SIM, identity verification is a process of monitoring users as they attempt to access the environment's resources and convert this knowledge into a numeric value (assertion value) that represents the system's confidence of the user's identity. This value is used to update the system's confidence as the user performs a known activity or otherwise. An integral component within SIM is a security policy (assertion policy) which uses the assertion value to decide whether the user can be trusted to access the resource or not. The assertion policy consists of a number of parameters whose collective values determine the nature of the policy applied for the environment. Although administrators can decide the type of policy that would best suit their specific environment by choosing appropriate values for these parameters, this research has conducted a number of trials to tune the system for strict, moderate or relaxed policies. The process of identifying user behaviour is done using an unsupervised clustering algorithm. A user activity is represented using a vector of user's identity, location and time, which the algorithm uses to cluster them to form event types. The clustering algorithm is based on probabilistic methods to identify recurrent event types or sequences of recurrent event types. Significant event types are recognised by the system every time a user performs one of them and hence, they are not used by the clustering algorithm. When an event type is detected, it will be converted into an assertion value and the identification process progresses.

In comparison to other identity verification methods, SIM offers a user-centric approach that is less intrusive, yet scalable and dynamic. The main limitation of SIM is the time it takes initially to build the system's knowledge of its users.

REFERENCES

[1] R. Roman, P. Najera and J. Lopez, "Securing the Internet of Things," *Computer*, Vol. 44, No. 9, 2011, pp. 51–58.

[2] Gubbi, J., et al., Internet of Things (IoT): A vision, architectural elements, and future directions. Future Generation Computer Systems, 2013.

[3] F. Mattern, C. Flörkemeier: From the Internet of Computers to the Internet of Things. Informatik-Spektrum, 2010, 33(2).

[4] L. Atzori, A. Iera and G. Morabito: The Internet of Things: A Survey, Elsevier Computer Networks, 2010.

[5] Alabady, S., Al-Turjman, F., and Din, S. "A Novel Security Model for Cooperative Virtual Networks in the IoT Era", *Springer International Journal of Parallel Programming*, 2018. DOI: 10.1007/s10766-018-0580-z.

[6] Jain, A., Bolle, R. and Pankanti, S., 1999, Biometrics: Personal Identification in Networked Society, USA: Kluwer Academic Publishers.

[7] A. Al-Fagih, F. Al-Turjman, W. Alsalih and H. Hassanein: A priced public sensing framework for heterogeneous IoT architectures, IEEE Transactions on Emerging Topics in Computing, vol. 1, no. 1, pp. 133–147, June 2013.

[8] F. Al-Turjman, A. Alfagih, W. Alsalih, and H. Hassanein: A delay-tolerant framework for integrated RSNs in IoT, Elsevier: Computer Communications Journal, vol. 36, no. 9, pp. 998–1010, May, 2013.

[9] Matyas, V. Jr, and Riha, Z., 2000, Biometric authentication systems, Tech. Rep. FIMU-RS-2000–08, Faculty of Informatics, Masaryk University, FI MU Report Series, Available from: http://www.fi.muni.cz/informatics/reports/files/older/FIMU-RS-2000-08.pdf.

[10] Gamboa, H., and Fred, A., 2004, A behavioral biometric system based on human-computer interaction, Proc. SPIE 5404, Biometric Technology for Human Identification.

[11] Barton, T., Basney, J., Freeman, T., Scavo, T., Siebenlist, F., Welch, V, Ananthakrishnan, R., Baker, B., Goode, M., and Keahey, K., 2006, Identity Federation and Attribute-based Authorization through the Globus Toolkit, Shibboleth, Gridshib, and MyProxy, 5th Annual PKI R&D Workshop, April 2006.

[12] Al-Turjman, F., and Alturjman, S. "Confidential Smart-Sensing Framework in the IoT Era", *The Springer Journal of Supercomputing*, 74.10(2018): 5187–5198.

[13] N. Chamberlin, Brief Overview of Single Sign-On Technology, Government Information Technology, 2000.

[14] Radha, V. and D.H. Reddy, A Survey on Single Sign-On Techniques. Procedia Technology. 2012. 4(0): p. 134–139.

[15] Hyun-Kyung Oh, Seung-Hun Jin: The Security Limitations of SSO in OpenID", 10th International Conference on Advanced Communication Technology, vol. 3, p. 1608–1611, 2008.

[16] Xu, Wei, V. N. Venkatakrishnan, R. Sekar, and I. V. Ramakrishnan. A Framework for Building Privacy-Conscious Composite Web Services. Tech. Stony Brook University. Web. June-July 2010.

[17] Wu Kaixing, Yu Xiaolin, A Model of Unite-Authentication Single Sign-On Based on SAML Underlying Web, In proceeding of: Information and Computing Science. ICIC '09. Second International Conference on, Volume: 2, 2009.

[18] Altmann, J. and Sampath, R., 2006, UNIQuE: A User-Centric Framework for Network Identity Management, Network Operations and Management Symposium, NOMS 3–7 April 2006. 10th IEEE/IFIP, 495–506.

[19] Al-Turjman, F., and Alturjman, S. "Context-sensitive Access in Industrial Internet of Things (IIoT) Healthcare Applications", *IEEE Transactions on Industrial Informatics*, 14.6(2018): 2736–2744.

[20] DEY, A; (2001), "Understanding of Context and Context-Awareness". Personal and Ubiquitous Computing, Vol.5, No.1, pp. (4–7)

[21] Al-Muhtadi, J., Hill, R., Campbell, R. and Mickunas, M. D. (2006), "Context and Location-Aware Encryption for Pervasive Computing Environments". Pervasive Computing and Communications Workshops,Percom Workshops 2006. Fourth Annual IEEE International Conference.

[22] Kulkarni, D. and Tripathi, A., 2008, Context-Aware Role-based Access Control in Pervasive Computing Systems, Proceedings of the 13th ACM symposium on Access control models and technologies, ACM New York, NY, USA, 113–122.

[23] Bubeck, S. and von Luxburg, U., 2009, Nearest Neighbour Clustering: A Baseline Method for Consistent Clustering with Arbitrary Objective Functions, Journal of Machine Learning Research, 10 (June 2009), 657–698.

[24] S. Oteafy, F. Al-Turjman, and H. Hassanein, "Pruned Adaptive Routing in the Heterogeneous Internet of Things", *In Proc. of the IEEE International Global Communications Conf. (GLOBECOM'12)*, Anaheim, California, 2012, pp. 232–237.

5

CONFIDENTIAL SENSING IN THE IoT ERA[1]

Contents

---

[1] Previously published in F. Al-Turjman, and S. Alturjman, "Confidential Smart-Sensing Framework in the IoT Era", *The Springer Journal of Supercomputing*, 2018. DOI. 10.1007/s11227-018-2524-1.

Abstract

With the revolution of the Internet technology, smart-sensing applications and the Internet of Things (IoT) are coupled in critical missions. Wireless Sensor Networks (WSNs), for example, present the main enabling technology in IoT architectures and extend the spectrum of its smart applications. However, this technology has limited resources and suffer from several vulnerabilities and security issues. Since the wireless networks used by this technology are deployed in open areas, several challenges are faced by the service provider in terms of privacy and the quality of service. Encryption can be a good solution to preserve confidentiality and privacy but it raises serious problems concerning time latency and performance. In this paper we propose agile framework that enables authentication, confidentiality and integrity while collecting the sensed data by using elliptic curve cryptography.

Keywords

Data delivery, Internet of Things (IoT), Wireless Networks, Big data

I. INTRODUCTION

The Wireless Sensor Network (WSN) shall handle remote monitoring and control of a large number of sensing and monitoring devices under variable density and mobility conditions. The network should operate under the constraints of low energy consumption and low power transmission, using reliable and cost effective networking techniques [1]. WSN has been invested in the Internet of Things (IoT) as a fundamental technology. The end user can thus experience potential threats caused by the wireless network nature. Several attacks have been raised to compromise the security of these networks. In order to avoid these attacks several studies have been conducted such as those in [7, 8] to offer more efficient solutions against these attacks. However, all these attempts are independent and can't be integrated together in one comprehensive system. Accordingly, authors in [9] have implemented an elliptic curve cryptosystem for WSNs, to reduce the cost while maintaining a good level of security. Authors in [10] also proposed a mechanism for clustered WSNs secure access based on the Elliptic Curve Cryptography (ECC). But these attempts are more interested in secure communications among nodes rather than the whole information system that allows end-users to recover the collected data confidentiality. Authors in [11] proposed the initial architecture for Medical Wireless Sensor Networks (MWSN) using encryption and emergency access control. Authors in [12] propose a secure healthcare framework to assure privacy of the patients. Nevertheless, there is not that much work in this area, and the existing model does not cover all security aspects.

On the other hand, cloud-based solutions have emerged to cover this area. The Cloud Computing [13] enables users of sensor networks to fully utilize wireless technologies in storing, sharing and retrieving the collected data by sensors anytime and anywhere. Despite of this, the use of cloud services is still very limited due to security and user privacy issues. Several attempts have been performed to offer an adequate solution to this problem [14–16].

Thus, we aim to build a secure platform, that secure the delivery process of the sensed data, while assuring authentication, confidentiality, integrity and privacy. We propose the use of cryptosystem based on elliptic curves [3]. Elliptical curve cryptography (ECC) [4] is a public key encryption technique based on elliptic curve theory that can be used to create faster, smaller, and more efficient cryptographic keys. According to literature [5, 6], ECC can yield a level of security with a 164-bit key only unlike other systems like RSA which requires a 1024-bit key.

Our proposed framework consists of authentication, confidentiality and a sharing scheme to secure the WSN. We discuss also some attacks in IoT-based environments, and propose a few security measures and detection techniques to prevent and secure the WSN from these attacks. We simulate our framework, assess and discuss the simulation results and security issues.

The rest of the paper is organized as follows. In Section II, we present our proposed model and we describe the design aspects which can assure the efficiency of our architecture. Then, we discuss some additional security measurements and detection techniques to improve the proposed framework. In Section III, we present a case study related to medical monitoring solutions, and a possible implementation for our framework. Then, we analyze security issues and provide a performance evaluation results. In Section IV, we conclude this work.

II. BACKGROUND

WSNs suffer from severe weaknesses such as node's limited resources, low communication quality and exposure to attacks. Table 5.1 represents the main attacks against this kind of wireless communication technology in IoT. In order to overcome these limitations and avoid these attacks, we propose the use of Elliptic Curve Cryptography (ECC). This technique is based on the mathematic properties of elliptic curves [6]. Elliptic curves are applicable for encryption, digital signatures, and other security aspects. We use properties of elliptic curves to assure a secure data transmission and communication. It assures the following: 1) Integrity of request and data shared between different entities in the IoT model, 2) Confidentiality of data collected by sensor nodes and saved in the Cloud server, and 3) Privacy and authentication of users. In Table 5.2, we compare ECC and other cryptography-based algorithms' in terms of the required key size to achieve the same level of security.

Since ECC uses a small size key, it allows encryption of the transmitted data more quickly, and thus, encrypted data will be much

Table 5.1   Main Attacks in IoT-Based WSNs

ATTACKS	LAYER	ATTACK DESCRIPTION	SUGGESTED SOLUTIONS
Node replications	DataLink	A few nodes are captured to collect encryption keys, and then replicates the node to redirect requested data.	Authentication, encryption and integrity mechanisms.
Collision	DataLink	Compromised packets are sent to make inter-node confusion.	Using error-correction codes
Jamming	Physical	Occurs by emitting a wide range of radio frequency in wireless signals.	Using detection technique and mobile agents
Sybil attack	Network	Malicious nodes are used to forges a large number of fake identities in order to disrupt the network.	Authentication and access control mechanisms, key management and position verification.
Black hole	Network	A few nodes in the network are used to block the packets they receive instead of forwarding and then cause a denial of service.	Authentication and access control mechanisms
Sink hole attack	Network	The sink node is compromised and all packets are blocked.	Authentication and access control mechanisms

smaller. ECC is also used in digital signature generation. Table 5.3 shows the number of signatures generated by these algorithms per second [17].

The type of exchanged information in the WSN has different constraints and urgency in accordance with the content of the communicated packets. Thus, the way the security protocols are applied must match with the confidentiality required for that specific packet. This creates a need to classify different levels of communication before even relaying/broadcasting them at the WSN used in IoT. The most crucial part in any IoT-based WSN application nowadays is ensuring that the network supports an end-to-end encryption and authentication. Critical key points to be considered in small cells applications for guaranteed security and privacy protection are as follow: 1) personal data collection, which if limited to certain extent can significantly help in mitigating several security issues, 2) data and traffic analysis for WSN based applications requires information sharing, therefore service providers and the technology partners should come into an agreement for a secure data handling methods which assures the mobile user privacy protections by considering the de-identification

Table 5.2 Comparative Analysis for the Key Size in Well-Known Algorithms and ECC

	ECC			AES			RSA		
KEY SIZE (BITS)	SIGNATURE GENERATION (MS/OPS)	SIGNATURE VERIFICATION (MS/OPS)	ENC/DEC	KEY SIZE (BITS)	ENC/DEC	KEY SIZE (BITS)	SIGNATURE GENERATION (MS/OPS)	SIGNATURE VERIFICATION (MS/OPS)	ENC/DEC
160	0.37	1.63	0.01/0.13	128	0.04/0.12	1024	0.27	0.02	0.02/0.29
224	0.41	1.87	0.02/0.46	256	0.06/0.78	2048	1.24	0.03	0.03/1.24
256	0.64	2.21	0.04/0.75	512	0.07/1.06	3072	3.45	0.06	0.05/2.78
384	0.76	2.76	0.06/0.88	1024	0.12/1.34	7680	6.12	0.12	0.09/5.14
521	0.94	3.49	0.08/1.32	2048	0.09/2.16	15360	19.41	0.19	1.06/7.35

Table 5.3 Signatures Generated by RSA and ECC per Second

ALGORITHM	KEY SIZE	SIGNATURE/SECOND
ECC	256	9516.8
RSA	2048	1001.8

concept for example, 3) reliability of the WSN itself, encryptions and digital signature per user are also important aspects to be considered in this domain alike with the management protocols and physical security of the WSN, 4) human errors, which can elevate security risks and breaches, and thus, customized policies and procedures are required to mitigate the oversight issues, 5) lastly, transparency of the WSN usage/configuration assures the integrity of such systems in wireless technology domains and necessitates accountably clear policies with respect to the offloaded data security and privacy. In order to satisfy these security needs, we propose agile WSN architecture, that preserve authentication, confidentiality and security/privacy in IoT applications.

III. SYSTEM MODELS

This section will discuss the system model of WSN and related assumptions.

A. Network Model

Figure 5.1 illustrates the system model of WSN in which the user may use a sensor in a smart phone to sense and update the monitored system/region to the technical expert. The expert may access the current information of the system via a mobile device, such as PDA and laptop. Thus, the targeted WSN model has three actors, namely user, system expert, and the authentic-gateway. Since the user and experts share the confidential information over an Adhoc network, this paper thus interpolates an authentic-gateway system into the Ad-hoc network to provide proper user authentication, secure-session key sharing and minimum computation overload, user friendly and resisting the potential attack of a privileged - insider.

In addition, a smart PnP sensor like IEEE 1451.7 (incorporated with the sensor) is used to define the hardware and software transducer

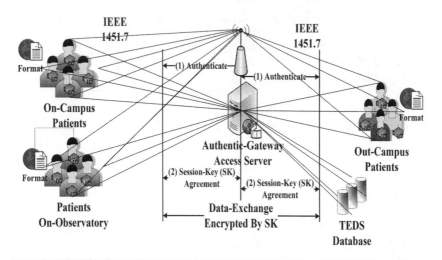

Figure 5.1   Network model of the assumed WSN.

interface (sensor or actuator). The specific objective of IEEE 1451.7 [16] is the sensor security that is interfaced between sensor and user, to achieve the features of mutual authentication and session-key establishment. IEEE 21451.7 is similar to the IEEE 1451.7, and thus this paper has preferred the IEEE 1451.7 for the achievement of sensor security operation and secure data (multimedia) transmission. To simplify the sensor descriptive characteristics, a smart sensor incorporated the Transducer Electronic Data Sheet (TEDS). The main purposes of this technology are: 1. To assist the host-device to identify the sensor related parameters; and 2. To provide the wireless connection to the host-device provided self-descriptive sensor parameters.

IV. CONFIDENTIAL WSN FRAMEWORK IN IoT

In this framework we assume a private cloud server for more privacy to avoid problems related to data collocation and virtualization.

As shown in Figure 5.2, our framework consists of the following main entities: 1) User (US), 2) Central Server (CS), 3) Sensor Nodes (SN), 4) Firewall, 5) Intrusion prevention server (IPS), and 6) the Private Cloud Server (PCS). Sensor Nodes collect data and send all these data to the local server. The firewall is responsible for monitoring the communications between the nodes and the central server

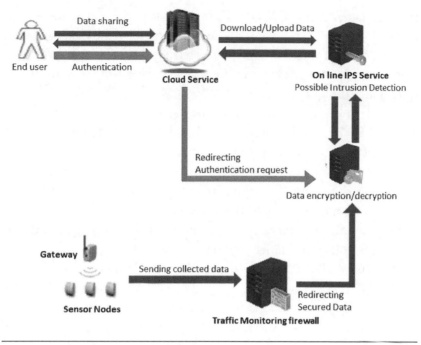

Figure 5.2 General architecture for the proposed framework.

and controls data traffics. The IPS controls data traffic and analyzes data in deep and compare different data's signatures to prevent external attacks. The local server encrypts data, authenticates users and ensures communication between the mobile device and the cloud server. The cloud server receives and stores the encrypted data. User after registration and authentication can download the encrypted data, decrypt and then conduct a secure data access. Consequently, our security scheme is divided into three main phases: 1) The user's authentication phase, 2) The downloading phase from PCS, and 3) The data sharing phase between users.

A. Authentication Phase

For authenticated data access, users should enter a proper long-term secret key into the smart phone system. Upon receiving the login-request, the authentic-gateway verifies the long-term secret key into the system database to execute the following operations: $N_i^* = H\left(M_{id} \oplus s_k \oplus S_{key}\right)$; Compare: $N_i^* = N_i$; compute: $H\left(M_{id}\right) \text{ and } CID_i = E_k[H\left(M_{id} \parallel M \parallel S_N\right]$; and eventually, generate $\{CID_i, C, T'\}$, then sends to the authentic-gateway node. Herein, M is a random nonce determined by the medical expert to establish secure session-key. Upon receiving the medical expert's message, the authentic-gateway executes the following tasks to authenticate his/her access-request:

Step 1: Initially, the authentic-gateway node validates the current access time T_c: validate whether $\left(T_c^{''} - T_c^{'}\right) \geq \Delta T_c$, if the expression holds, then the authentic-gateway node refuses the access-request and terminate the process. Otherwise, the authentic-gateway executes the further steps. Herein, $T_c^{''}$ is the current request-time of the authentic-gateway and ΔT_c is the delay time interval.

Step 2: After the successful validation, the authentic-gateway executes the following tasks: Compute: $D = D_J\left[M_{id}^* \parallel ID_{gw}^*\right]$ from the M_{id}^* and ID_{gw}^*; Compute: $H\left(M_{id}^*\right)$; Compute: $D_k\left[CID_i\right] = E_k\left[H\left(M_{id} \parallel M \parallel S_N\right)\right]$ from the $M_{id}^{'}, M$ and S_N. Compare $H\left(M_{id}^*\right) = H\left(M_{id}^{'}\right)$ and $ID_{gw}^* = ID_{gw}$; if the condition is satisfied, then the request is authentic; Otherwise, terminate the rest of the process; Compute $V_i = E_{SK_{gw}}\left[M_{id} \parallel S_N \parallel M \parallel T_c^{''}\right]$; generate the request-message $\{V_i, T_c^{''}\}$ and then sends the request-message to the nearby medical sensor / access point wherein the medical expert is available to access the patient info.

Step 3: Upon receiving the authentic-gateway message, the medical sensor / smart phone executes the following tasks: S_N verifies the time $T_c^{'''}$: validate whether $T_c^{''''} - T_c^{'''} \geq \Delta T_c$, if the validation is successful, the medical sensor node rejects the authentic request-message of the authentic-gateway and terminate the further process. Herein, $T_c^{''''}$ is the current execution time of the medical sensor node and ΔT_c is the delay time interval.

B. Downloading Phase

In this phase, the user enters the credentials into the authentic-gateway system and its related execution flows are as follows:

Step 1: The user chooses M_{id} and s_k and then he / she submits into the authentic-gateway node via secure channel.

Step 2: Upon receiving the requested data M_{id} and s_k, the authentic-gateway determines the followings: $C = E_J\left[M_{id} \parallel ID_{gw}\right]$ and $N_i = H\left(M_{id} \oplus s_k \oplus S_{key}\right)$.

Step 3: Thenceforth, the authentic-gateway provides a secure-ware to the user with the configuration of the following parameter $\{H(.), C, N_i, S_{key}\}$. Herein, S_{key} is a long-term gateway secret key that is securely bound between the entities.

C. Sharing Phase

This phase is called forth while U_{ser_i} wishes to update his / her shared data. The working procedure for secret-update is as follows:

Step 1: U_{ser_i} puts his / her shared data in to the terminal to enter his / her credentials, namely M_{id} and s_k.

Step 2: Upon the credentials validation and verification, the user out the operation of $N_i' = H\left(M_{id} \oplus s_k \oplus S_{key}\right)$ to compare with $N_i' = N_i$. If the comparison is successful, then the rest of the operation will be proceeded. Otherwise, the operation will be aborted.

Step 3: If validation is successful, U_{ser_i} is asked to enter a new secret-key s_k^{new}.

Step 4: Compute $N_i^{new} = H\left(M_{id} \oplus s_k^{new} \oplus S_{key}\right)$ to replace N_i with N_i^{new} for that user.

V. PERFORMANCE EVALUATION

We propose in this part an implementation of the proposed framework for the medical monitoring applications. We chose the medical field because data collected from patients are considered highly sensitive according to social, ethical and legal aspects of medical systems [18]. The purpose of our implementation is to secure the communications process between sensor nodes, the central server that manages, encrypts, and store data in a private cloud server, and medical staff who retrieve data from the Cloud, decrypts it, and use the information obtained as required.

In our implementation, we have deployed sensors nodes to collect data in real time. All these data are sent to the local server that will be in charge to sort, encrypt and organize in a local database. These data are stored in a private cloud server, this server allows us to preserve privacy and prevent several problems of virtualization and data collocation. Then each user will be able to log from any machine connected to the Internet (Mobil, Computer, Laptop…) to retrieve the data stored in the server.

For this it must first register in the system through their personal data (Pass, ID, Name, phone number), then it must issue an authentication request for each new connection attempt. This procedure will allow it to connect to the private cloud server safely and recover the encrypted data. Each user can decrypt the data using our confidentiality model that we have established beforehand.

A. Performance Matrix

To compare the performance of the proposed framework, the following three performance metrics are used:

1. Time: represents the time that the event will take until accomplished in the system and measured in milliseconds.
2. SKGR: Secret key generation rate measured in bits per channel per user.

While studying these performance metrics, we vary the following parameters:

3. Key size: In the proposed scheme, MU_i and S_N can determine a session-key size using the computation of $SK_{GS} = H\left(MID_i^* \parallel S_N \parallel X \parallel TS' \parallel TS^*\right)$; and thus, it indicates the access operation complexity in its execution.
4. SNR: Signal to noise ratio, which is used to measure the received signal quality.

B. Simulation Results

We present in this section, the results of the afore-described implementation for the proposed framework. We start with the authentication phase, which will allow users (or doctors) to authenticate safely while accessing the central server. Next we present the results of encryption and decryption of data according to our confidentiality scheme.

To recover the data collected by the nodes, each user must register and log in via a mobile application. Then, each user must enter personal data to be authenticated via the central server. The server checks the validity of the request and confirm the user identity. Results of

Table 5.4 Authentication Process Results

REGISTRATION:	
Login:	Fadi155
Password:	123456FA
SIGNATURE GENERATION:	
Name	Fadi Al-Turjman
Pass:	123456FA
ID:	00556677
Phone number:	01234567
signature:	68801224901208227004381526319392425483005131202835940388513534916420427370224638057252239606366642183
authentication request	(Fadi Al-Turjman, 123456FA, 00556677, 01234567)
Encrypted authentication request PKSC(NC,ID,NP):	C39BEC704F4365E18230D25FDBE7565F16D418A2D274056922A6C6 AE46 + Signature
RECEIVED REQUEST FROM USER:	
Encrypted authentication request	C39BEC704F4365E18230D25FDBE7565F16D418A2D274056922A6C6 AE46 + Signature
SIGNATURE VERIFICATION	
Computes new signature:	02835940388513534916420427370224638068801224901208227004381526319392425483005131 2
final verification:	1

the various steps of the aforementioned authentication phase in our framework are as shown in Table 5.4.

After receiving the session key from the central server, each user can connect to the private cloud server to retrieve the collected data. The collected data is encrypted by the central server and decrypted by the intended user using ECC scheme. Using our confidential framework, encryption and decryption of data is performed in a very fast way and without overloading the CPU.

Our proposed framework is powerful because it offers a satisfactory level of data security and can generate keys and signatures very quickly. Communication between the various described entities in our framework is secure as well. It is also very scalable because it allows multiple security policies and it adapts to any future WSN development. Thus, it allows flexible and fast interconnection and offers a sharing platform that allows users to exchange data securely.

The use of encryption assures a good level of data security. However, it can affect the performance of the proposed framework in terms of speed. Accordingly, we took this into consideration and we used ECC encryption based system. We also used lightweight and efficient processes during authentication, and key/signature generation to reduce the overall delay. Figure 5.3 shows the encryption process according to different keys' size and Figure 5.4 shows the authentication process and the signature verification process performance as well. Results of our different experiments show that our framework is light

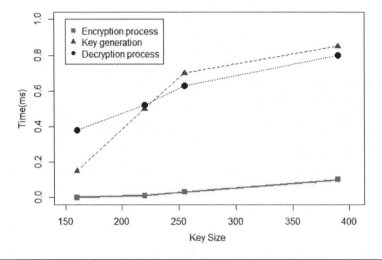

Figure 5.3 Encryption and decryption process.

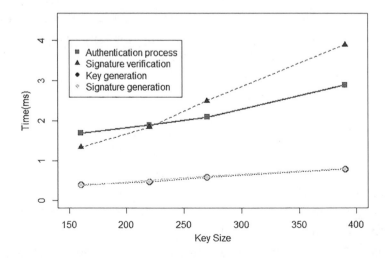

Figure 5.4 Authentication and signature generation process.

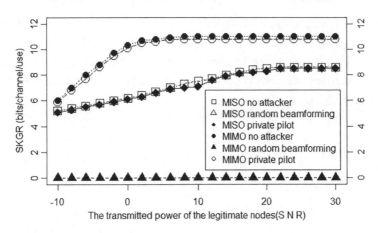

Figure 5.5   SKGR varies with SNR of transmitter.

and assures integrity and privacy. Since the proposed framework has less computation cost, it mitigates the execution time of the phase to improve the performance of the WSN in comparison with the other authentication schemes.

To verify the security and suitability of the proposed secret key generation (SKG) approach, we compare the SKG Rate (SKGR) of the proposed scheme based on private pilot with the approaches based on random beam-forming described in [20].

Figure 5.5 shows that the SKGRs based on the private pilot and reconstitution channels rise with the increasing of the SNR of the transmitter, which is close to the SKGR without attacker. However, the SKGR based on random beam-forming with public pilot described in [28] are almost equal to zero under the man-in-the-middle (MITM) attacks, which shows that the MITM attack is a serious threat to SKG.

VI. CONCLUSION

In this paper we propose a confidential framework in WSN to ensure integrity, confidentiality and privacy. We also use a private cloud computing for extended storage and computation resources, for instantaneous data access. Our framework assumes an elliptic curve cryptography model to assure data encrypting/decryption while collected from sensor nodes. After detailing the design aspects of the proposed framework, we discuss related security attacks in WSN/IoT environments and investigate security metrics and detection techniques. Furthermore, an implementation of the proposed framework for medical applications is performed. Significant results and observations have been reported.

In our future work, we aim at further improving our confidential framework in terms of complexity and access time for better quality of experience. In addition, we plan to implement our platform in other fields related to the Cloud in practice for large-scale applications.

REFERENCES

[1] Yick, J., Mukherjee, B., Ghosal, D.: 'Wireless sensor network survey'. Computer Networks 52, 2008, pp. 2292–2330.

[2] Boginski, V. L., Commander C. W., Pardalos, P. M.: 'Sensors: theory, algorithms, and applications'. Vol. 61 of Springer Optimization and Its Applications, Springer-Verlag, New York, 2011.

[3] Enge: 'Elliptic Curves and their Applications to Cryptography, an Introduction'. 1999.

[4] Koblitz, N., Menezes, A., Vanstone, S.: 'The State of Elliptic Curve Cryptography'. Towards a Quarter-Century of Public Key Cryptography, 2000, pp. 103–123.

[5] Gura, N., Patel, A., Wander, A., Eberle, H., Chang Shantz, S.: 'Comparing Elliptic Curve Cryptography and RSA on 8-bit CPUs'. Lecture Notes in Computer Science Volume 3156, 2004, pp. 119–132.

[6] Maletsky, K.: 'RSA vs ECC Comparison for Embedded Systems'. Senior Product Line Director, Security ICs, 2015.

[7] Yang, C. L., Tarng, W., Rong Hsieh, K., Chen, M.: 'A Security Mechanism for Clustered Wireless Sensor Networks Based on Elliptic Curve Cryptography'. Intelligent Internet Systems, 2010.

[8] Debnath, A., Singaravelu, P., Verma, S.: 'Privacy in wireless sensor networks using ring signature'. Journal of King Saud University – Computer and Information Sciences, 2014.

[9] Boyle, D. E., Newe, T.: 'On the implementation and evaluation of an elliptic curve based cryptosystem for Java enabled Wireless Sensor Networks'. Sensors and Actuators A 156, 2009, pp. 394–405.

[10] Sarigiannidis, P. Karapistoli, E., Economides, A.: 'Detecting Sybil attacks in wireless sensor networks using UWB ranging-based information', Expert Systems with Applications 42, 2015, pp. 7560–7572.

[11] Lounisa, A., Hadjidja, A., Bouabdallaha, A., Challala, Y.: 'Healing on the Cloud: Secure Cloud Architecture for Medical Wireless Sensor Networks'. Future Generation Computer Systems, 2015.

[12] A Cloud-based Healthcare Framework for Security and Patients Data Privacy Using Wireless Body Area Networks'. The 9th International Conference on Future Networks and Communications (FNC'14)/The 11th International Conference on Mobile Systems and Pervasive Computing (MobiSPC'14)/Affiliated Workshops, Volume 34, 2014, pp. 511–517.

[13] Mell, P., Grance, T.: 'The NIST Definition of Cloud Computing', Special Publication, 2011.

[14] Buyya, R., Yeo, C.S., Venugopal, S., Broberg, J., Brandic, I.: 'Cloud computing and emerging it platforms: vision, hype, and reality for delivering computing as the 5th utility', Future Generation Computer Systems 25, 2009, pp. 599–616.

[15] Zkik, K., Orhanou, G., El Hajji, S.: 'Secure scheme on mobile multi cloud computing based on homomorphic encryption', Proceeding of the International Conference on Engineering & MIS (ICEMIS), 2016, pp. 1–7.

[16] Zkik, K., Orhanou, G., El Hajji, S.: 'Secure Mobile Multi Cloud Architecture for Authentication and Data Storage', International Journal of Cloud Applications and Computing (IJCAC), vol. 7, issue 2, 2017, pp. 62–76.

[17] Sullivan, N.: 'ECDSA: The digital signature algorithm of a better internet', 2014.

[18] Rodrigues, J. J., de la Torre, I., Fern´andez, G., L´opez- Coronado, M.: 'Analysis of the security and privacy requirements of cloud-based electronic health records systems'. J Med Internet Res 15 (8), 2013.

[19] L. Cheng, W. Li and D. Ma, "Secret Key Generation via Random Beamforming in Stationary Environment", *International Conference on Wireless Communications & Signal Processing (WCSP)*, 2015, pp. 1–5.

[20] Al-Turjman, F., "Fog-based Caching in Software-Defined Information-Centric Networks", *Elsevier Computers & Electrical Engineering Journal*, vol. 69, issue 1, 2018, pp. 54–67.

6

IoT Smart Homes and Security Issues: An Overview

FADI AL-TURJMAN AND CHADI ALTRJMAN

Contents

Abstract

Homes' owners currently lack the means to gather quantitative multi-modal data about the long-term behaviors of residents, appliances, and resources in their home environments. Smart home is a multi-modal platform of heterogeneous sensors for behavior monitoring in residential environments that aims to overcome this major aspect through using the inherently cost-efficient and scalable technologies of Internet of Things (IoT). One of the IoT paradigm key tasks is to help to bring the next-generation low-power wireless networking and sensing technologies from the lab to the field by applying them in real-world environments. In this article we highlight the smart home system requirements, architectures, practical challenges, as well as of the design and deployment aspects.

Keywords

Smart homes, IoT, Cloud Computing

I. INTRODUCTION

A smart home is an application enabled by ubiquitous computing in which the home environment is monitored by ambient intelligence to provide context-aware services and facilitate remote home control [1]. Furthermore, it is considered as a combination of several enabling technologies such as sensors, multimedia devices, communication protocols, and systems. From a different scope, a smart home is merely a residence equipped with different Internet-connected devices to remotely monitor and manage the appliances and systems installed in the home such as lighting and heating, to just mention a few. Such a smart residence would be useful in managing the daily lives of the inhabitants. With the recent developments in the Information and Communication Technologies (ICT) and the reduction in the costs of low-powered electronics, a new technology has drawn the attention of the research community, namely the Internet of things (IoT).

IoT is a revolutionizing technology tends to connect the entire world by connecting physical smart devices used for sensing, processing, and actuating [2-3]. By integrating the Machine to Machine (M2M) communication technologies with the smart devices, these devices can connect and interact without any human intervention. As a result, IoT is believed to enable a fully conductive environment that can influence the life of the society in different aspects such as every day's activities of the individuals, business and economy applications, healthcare applications, energy applications, traffic and road controlling and even in political systems, to just mention a few. Moreover, the "Things" are merely the devices and objects connected to a common interface with the ability to communicate with each other. By integrating the three core components of the IoT, namely the internet, the things, and the connectivity, the value of IoT is to close the gap between both the physical and the digital worlds in the self-reinforcing and self-improving systems. The concept of smart homes is considered as an IoT- based application enabled by connecting the home appliances to the internet. to prevent home system's main goal is to provide Security, monitoring and controlling for all devices in homes over a cloud.

To achieve security, the system detects any threats in the home such as gas leaks, water leaks, and fires, then, it is alarming the residents to prevent any losses in lives or properties. In addition, the system provides instant detection for any robberies happening. The controller manages all the devices installed in the home and it can remotely control these devices with the aid of smartphones. In addition, the system is compatible with all kinds of devices with the ability to manage their running time. Amazon Web Server has built IoT specific services, such as AWS Greengrass, and AWS IoT. These services help people to collect and send data to the cloud, to load and analyze data, and to manage devices. AWS IoT is a managed cloud platform which allows the connected devices to easily and securely interact with each other and with cloud applications. AWS IoT is a managed service built for the purpose of connecting the devices to each other and to the cloud. Moreover, it can handle billions of devices and trillions of

Table 6.1   Acronyms Used in this Article

ACRONYM	DESCRIPTION
AWS	Amazon Web Server
PIR	Passive Infrared sensor
MW	Microwave
LED	Light Emitting Diode
UV	Ultraviolet
IoT	Internet of Things
AC	Air Conditioner
VRF	Variable Refrigerant Flow
VRV	Variable Refrigerant Volume
UPB	Universal Powerline Bus
BLE	Bluetooth Low Energy
ISM	The industrial, scientific and medical radio band
HVAC	Heating, ventilation, and air conditioning
P2P	Peer-to-Peer
WSDL	The Web Services Description Language
SaaS	Software as a service
PaaS	Platform as a service
IaaS	Infrastructure as a Service
OSGi	Open Services Gateway Initiative
XML	Extensible Markup Language
SOAP	Simple Object Access Protocol
UDDI	Universal Description, Discovery, and Integration
HGW	Home Gateway
WLAN	wireless local area network
ULD	User-friendly Location Discovery
OEM	Original Equipment Manufacturer

messages, with the ability to reliably and securely process and route these messages to the AWS endpoints and to other devices.

In this paper, we will overview the IoT-based smart homes, the sensor types that can be deployed, the enabling standards and the cloud architecture of a smart home. The rest of the paper is constructed as follows. In Section II, we will overview the different types of sensors used in smart homes. Section III discusses the communicating protocols and how they work. In Section IV, the common applications that can be deployed in smart homes are proposed. Section V discusses the cloud architecture for smart homes, and the paper is concluded in Section VIII. For more readability, we summarize our used abbreviations in this article in Table 6.1.

II. SMART HOMES' SENSORS

A sensor is an electronic component, module, or subsystem whose purpose is to detect events changing in its environment and send the information to other electronic devices, mostly a computer processor. A sensor detects and responds to some type of input from the physical environment. The output is generally a signal that is converted to human-readable display at the sensor location or transmitted electronically over a network for reading or further processing. Sensors in smart homes can be classified as follows:

A. Motion Sensors

This type is used in numerous applications including home security lights, automatic doors, and bathroom fixtures, that transmit some type of energy, such as microwaves, ultrasonic waves, or light beams, to mention a few. In addition, these sensors can detect when the energy flow is interrupted by something entering its path. In the following, we list a number of these applications:

Window & door control - This type of sensors monitors the doors and windows of the smart home and informs the owner about the people who entered or left his house. Moreover, these sensors can save the energy consumption by switching the lights on and off when someone opens or closes the doors. The window-deployed sensors can be considered as the first line of defense from home break-ins.

Further, some of these sensors detect any intruders trying to enter the house and alert the inhabitants. As these sensors are wirelessly connected to the internet, the owner of the house can receive notifications on his smartphone and call for any help needed.

Glass Break Sensors- Once a window is broken the smart house sensor sends a signal back to the smart home panel.

Door contacts- Once armed, a signal is sent to the smart home control panel, if the door is opened.

Video doorbell- this device is protecting the properties as working as a robbery hindrance sensor. Further, it monitors anything behind the doors. This enables the house owner to know anybody approaches his doors, whether he is staying at home or away from it.

Passive Infrared (PIR) - This type of sensors detects infrared energy, so it can recognize any human near to it by sensing his body heat. These sensors are widely used in home security-based applications by detecting the heat and movement, then they create a protective grid. For instance, if a moving object blocks different zones, the Infrared levels change and the sensors detect the movement and activate the protection grid. In [4], an application used for human localization in an indoor environment. This application uses the PIR sensors.

MircoWave (MW) - This kind of sensors sends out microwave pulses to measure the reflection of moving objects. MW sensors have an advantage over the infrared sensors, that it can cover more area. However, they are expensive and vulnerable to electrical interference. One of the common applications of Microwave sensors is human identification in smart homes as proposed in [5].

Dual-Technology Motion Sensors - These sensors use multiple technologies such as passive infrared (PIR), and Microwave (MW) sensors, to monitor a specific area. In this type, both sensors must detect a signal to trigger the alarm or to activate a specific process. This can reduce the instances of false alarms.

Area Reflective Type - This type of sensors has a LED that emits Infrared rays. The sensor detects an object if these radiated rays are reflected off it. Hence, the sensor measures the distance to the object and recognizes if the object is within the designated area

Ultrasonic sensors - They send out pulses of ultrasonic waves and measures the reflection off a moving object.

Vibration - Detects vibration. There are two main sensor types in this category – the accelerometer and the piezoelectric device.

Photonic Sensors - detects the presence of visible light, infrared transmission (IR), and/or ultraviolet (UV) energy.

B. Physical Sensors

This type is used in numerous applications as listed below:

Temperature - Installed in beds and in some chairs to detect user location based on body temperature.

Tap sensor - this type of sensors uses Infrared energy to sense the presence of an object. For instance, if the user puts his hands in front of the tap, the Infrared sensor detects them and sends a signal to the solenoid valve to allow the flow of water. On the other hand, when the user moves his hands away from the tap, the sensor sends a signal to the valve to immediately terminate the water flow. As a result, this can help in saving water.

Humidity sensor - These sensors can be installed in bathrooms to detect any rapid increase in humidity levels which are defined by the users. Further, the sensor sends these data to the Home Gateway (HGW).

Security - With an unlimited number of security sensors and settings, the residents can have full protection at home. While they are not home, they can start a pilot program to give an impression as if you are at home. Moreover, these sensors can Follow the intruders, switch off/on your lights depending on the motion. Also, while there is nobody at home, the security sensors can follow the kids at home to ensure their safety.

Flood - This kind of sensors Protects your home and workplace from floods. With an early warning system of any water leakage within the house, the related valve can be shut down.

Gas leak - With the help of gas detectors, the sensor gets an alert for slightest gas leaks. In addition, it will shut down the related valve. Thus, it will prevent any trouble that would occur due to such gas leakage.

Panic button - In emergency cases, this sensor would text or email your relatives or security institutions to send the help needed.

Shake sensor - In case of earthquakes, early warning system is activated. The system will act as your predetermined actions.

Irrigation - these sensors can check your irrigation valves at your garden, create your own irrigation program, or determine automatic irrigation scenarios according to the weather.

Curtain control - These sensors can remotely Control your curtains or canopy. You can also protect your home from extreme weather conditions by prescheduled scenarios.

Multimedia control - Control your home theater from this distance with your smartphone.

Ventilation - Split AC, VRF, VRV and any kind of ventilation system at your disposal, program and control it.

TV control - Change channels, increase-decrease volume using smartphones.

Smart plug - Control your plugged devices remotely. Measure energy consumption, automatically shut down devices even when you are not at home.

C. Chemical Sensors

This type is used in numerous applications as listed below:

Fire/CO detection - Fire is considered as the main reason for property damage. For years, the humble fire detector has been used for detecting any smoke in the homes. Nevertheless, there are many pollutants threatening the air quality inside the smart homes, resulting in harming the residents. One of these dangerous pollutants is Carbon monoxide (CO) which can be sensed by a CO detector. This device can be employed in emergency monitoring service to sense this odorless gas. The new versions of these sensors have many features. one of them is that they can't only monitor pollutants such as dust, soot, pollen, and particulates, but also, they monitor the overall air quality such as the humidity, and the air stakes. Moreover, the customer can receive attractive discounts from insurance companies, when he uses these sensors.

D. Leak/Moisture Detection

One of the main reasons for home damages is water flooding. For instance, if a water pipe is broken in the house and water flows from it continuously for many hours, this will destroy the devices and appliances in the home. A common solution for this is to deploy a moisture detection sensor to alarm the house owner if there is a water leak in it. These sensors can be placed near to water heaters, dishwashers, refrigerators, sinks and sump pumps. Further, if the sensor detects any unwanted moisture, the smart system can send a notification to the house owner about the problem, to find a solution for it quickly.

E. Remote Sensors

Smart garage door - The wi-fi connected smart garage door is a device helps the house owner in cases of forgetting closing the garage door. It enables the user to remotely control the garage door via a smartphone.

Smart thermostat -This device allows the user to remotely control the heating and cooling system within the house. Furthermore, it can control the humidity and adjust the temperature of the house based on the user's behavior inside the house. Moreover, these sensors can activate the energy saving mode when there are no users in the house. Applying cognitive technology to these sensors is a paving way for a smart home to be able to learn more about the residents and their temperature preferences.

F. Biosensors

Alert Z-Wave Smoke Detector and CO Alarm - When it detects smoke or carbon monoxide, it sounds an alarm and sends an alert to your phone. It can also contact any other number of people, including the fire department, if you're away. The First Alert detector, namely Albert Z-Wave Smoke Detector, is a great option that works with your SmartThings (or other Z-wave) hub.

Smoke - These sensors are used for detecting smoke. they can be installed in the ceiling of each partition. If a user is smoking, his location is identified based on a signal sent by this sensor.

III. PROTOCOLS FOR SMART HOME DEVICES

Protocols are all about how signals are sent from one device to another in order to trigger an action, such as turning the lights on and off.

There are a wide variety of technology platforms, or protocols, on which a smart home can be built. Each one is, essentially, its own language. Each language is related to various connected devices. In addition, it instructs them to perform a function. Choosing a smart home protocol can be a tricky business. It is better to choose a protocol that can support a large number of devices, and offer the best possible device interoperability, i.e. the ability for devices to talk to each other. Nevertheless, there are many factors to consider while choosing a protocol, such as power consumption, bandwidth and, of course, cost [6].

A. How Home Automation Protocols Work

Home automation protocols are considered as the communication frame used by the smart home devices to interact with each other, due to the vital role of communication to achieve the required home automation. It involves automatic controlling of all electrical or electronic devices in homes or even remotely through wireless communication. Centralized control of lighting system, air conditioning and heating, audio/video systems, security systems, kitchen appliances and all other equipment used in home systems are enabled by this kind of protocols. Moreover, implementation of the home automation depends on the type of controls like wired or wireless.

B. Differences Between Home Automation Protocols

There are mainly three types of home automation systems. Powerline Based Home Automation: This automation is inexpensive and doesn't require additional cables to transfer the information. Instead, it uses existing power lines to transfer the data. However, this system involves a large complexity and necessitates additional converter circuits and devices. Wired protocols such as UPB and X10, use the wiring existed in homes for communication. Although these protocols are reliable, they are slow and can face several difficulties to encrypt. On the other hand, wireless protocols usually tend to be faster and more congenial with many other devices. Wireless home automation protocols such as Z-Wave, ZigBee, Wi-Fi, Thread, and Bluetooth, are easier to be secured because they can communicate without using any power lines. Further, there are some protocols utilize both wired and wireless communication such as C-Bus and Insteon. As a result, they avail from the features provided by both wired and wireless technologies. Moreover, protocol's compatibility with smart devices is a measure of how many smart devices that the protocol can be easily employed with [7].

Z-Wave – It is seen as a wireless protocol working in the radio frequency band and used by many home automation devices to communicate with each other. Z-Wave runs on the 908.42 MHz frequency band. This frequency band is a much lower than the one used by most household wireless products (2.4 GHz). As a result, it is not affected by their interference and there are no traffic jams in this band. One significant advantage of Z-Wave is the interoperability of the devices connected to it which means that the devices communicate to all other Z-Wave devices, regardless of type, version or brand. Further, the interoperability is backward- and forward-compatible in the Z-Wave ecosystem.

How It Works - In order to set up and manage the home automation network, Z-Wave uses one central hub. In addition, the user can easily add any smart home devices and manage them using the Z-Wave protocol once the network is installed. [8].

Z-Wave Compatibility - With more than 1700 certified Z-Wave devices available in the world, it is easy for the homeowners to select from a plenty of options for automating their homes. In addition, Z-Wave devices can be easily set up and use and consume less energy.

Benefits of Z-Wave - As mentioned before, the Z-wave uses a frequency band which is much lower than the frequencies used for most of the other wireless devices. As a result, there is a lower chance to interfere with other frequencies which are crowded with many competing devices. So the Z- wave has a more efficient communication with home automation devices. Further, regardless the type or the version of the devices, they can freely interact with each other. This works also for the continuously evolving product. So, every time these products get updated, the users won't start with a new hub. This has a special importance nowadays where everything is updating quickly.

ZigBee – It is a popular wireless home automation protocols which is the same as Z-wave protocols in many ways. ZigBee was originally developed to be commercially used. However, it is now utilized as a standard protocol for many automation applications in both residential and commercial environments.

How It Works - Like the Z-wave protocol, The ZigBee protocol uses radio frequency to communicate. Furthermore, The ZigBee protocol is a low- cost, low- powered technology, which means that the battery- operated devices in a ZigBee will enjoy a long life. Moreover, it uses a mesh network which enables a larger ranged and a faster communication. Also, One ZigBee hub can be used by multiple devices which is suitable for automating home appliances.

ZigBee Compatibility - Recently, the ZigBee technology is considered as an open environment for developers to design many applications that can work with it. As a result, currently, there are more than 1200 products certified and compatible with a ZigBee hub. Early on, interoperability of the connected devices, which are made by several manufacturers, was a big challenge facing Zig-Bee. However, the recently released versions tend to have a better operability, regardless of their manufacturer or version.

Benefits of ZigBee - ZigBee has a low power usage, so the user can use a ZigBee device without changing the batteries for several years. Further, enabling ZigBee at home automation offers a green power feature which makes it unnecessary to use any batteries. For the security issues, Using the same level of encryption that most of the financial institutions use, it is seen as one of the most protocols having a high-security level. This high security will cover the ZigBee network, devices, and the information conveyed by it. Furthermore, ZigBee home automation is highly customizable and can be perfectly used by techies.

Insteon - The Insteon protocol is hybrid technology, which uses both wireless and wired communication technologies. This makes it unique to be added to the field of home automation.

How It Works - Insteon home automation runs on a patented dual-mesh network that uses both wired and wireless communication to overwhelm general difficulties facing each type while operating solely. An Insteon hub connects with Insteon-compatible devices. This creates the ability to control the user's smart home via smartphones.

Insteon Compatibility - With more than 200 Insteon-compatible home automation devices available commercially, Insteon tends to rely more on them, unlike many other protocols. This creates a restricted compatibility with smart devices made by other manufacturers. On the other hand, Insteon- compatible devices have a feature to be forward and backward compatible, meaning that they can work well with both new and old versions of devices.

Benefits of Insteon - Regardless of what the user's level of technical prowess is, he can run Insteon home automation with no running problems. the user needs only to know how a smartphone is used and the Insteon protocol is easy to be managed. Moreover, as the user turns any Insteon-compatible devices, they can be automatically added to the network. This makes them speedy-setup devices and the user faces no troubles with connecting them. In addition, the Insteon protocol shows a notable feature regarding the network size, the network is not limited, and it can be as large as the user desires. This is because that the network is a dual-mesh and one Insteon hub is able to work with hundreds of devices across a broad range, without facing any problems.

Bluetooth - It is considered as the core communication frame for hundreds of smart products nowadays. Having a higher data bandwidth than ZigBee and Z-Wave, made it suitable for home automation applications. However, its bandwidth is lower than the Wi-Fi, it consumes far less power than Wi-Fi. Bluetooth Smart, or Bluetooth Low Energy (BLE), is an energy efficient version of Bluetooth wireless technology often seen in smartphones, and ideal for use when the smartphone is connected to a headset. Because it is an energy efficient protocol and compatible with the existing smart devices, it is easy for developers and OEMs to create solutions that can immediately be added to the existing systems. Moreover, although it uses different channels, BLE operates in the already crowded 2.400 GHz-2.4835 GHz ISM band and the required data compression diminishes the audio quality. Further, Ble can be used within the home due to its low energy usage which can extend the battery life of the devices [9].

How It Works - Bluetooth is wireless home automation protocol that uses Radio frequencies for communication. Containing a computer chip with a Bluetooth radio and a software, makes it simple for various devices to interact with each other. Moreover, the users can use one primary Bluetooth hub for controlling all the devices connected to his home automation network.

Bluetooth Compatibility - with the ability to connect Bluetooth-enabled devices with each other, hundreds of Bluetooth-compatible products commercially exist now. Nevertheless, other non-Bluetooth devices can't be added to such a Bluetooth hub. Moreover, Bluetooth connections have a limiter ranged connection, as a result, many congenial devices can fail to work in such network because of the limited range of its communication.

Benefits of Bluetooth - Bluetooth is one of the fastest-growing sectors of home automation and it is included in many devices. This is due to that its home automation products are desired in many applications and consume little running power. Resulting in reducing the carbon footprint.

UPB – It stands for universal power line bus. It is a home automation protocol that uses the wired communication technology. UPB was released in 1999 and is considered as one of the most technically advanced protocols.

How It Works - Based on the X10 standard, the UPB home automation protocol uses the power lines for enabling communication between different devices. UPB devices can connect to the network using two enabling devices, a central home controller, which is manually set up by the user, and links for each device connected to the network. Further, UPB protocol doesn't require too much technical savvy for the users to set up and run.

UPB Compatibility - Compared to the other protocols, UPB has fewer home automation products available nowadays. One reason for that is that UPB is difficult to be combined with any wireless protocols. On contrast, there are around 150 commercial UPB- compatible products existing in the market.

Benefits of UPB - For the reliability issues, UPB is one of the most reliable protocols. This is because it is hard-wired into the power lines, so a limited interference is experienced compared to wireless home automation technologies.

THREAD - It is a new wireless protocol for smart household devices. The Thread Group was formed in July 2014 by seven founding members, including Google's Nest Labs and Samsung Electronics. More than 250 devices can be connected to a Thread network. Further, because most of the devices meant to be connected to the network are battery-operated, it's very frugal on power. Using the same frequency and radio chips as ZigBee, Thread is intended to provide a reliable low-powered, self-healing, and secure network that makes it simple for people to connect more than 250 devices in the home to each other. Moreover, they can be connected to the Cloud for ubiquitous access. Nest Learning Thermostat and Nest Protect are already using a version of Thread, and more products are supposed to enter the market soon.

5G- 5th Generation is the up and coming era of versatile advancements and is being intended to give more prominent ability to remote systems, offer more noteworthy unwavering quality, and convey to a great degree of speeds, empowering creative new administrations crosswise over various parts and facilities at home. With 5G, the Information and Communication Technologies will create new administrations at the minimal effort not just concentrating on giving a consistent and productive correspondence capacity as in the past, but additionally attempting to truly enhance the way we connect among ourselves, while enhancing our lives. Correspondence administrations will be free for the end client as a rule and will adapt every one of those smart homes applications, machines and things, that will be offered as an administration, hence permitting the move towards a more genuine Information-situated society.

LiFi- The potential spectrum crisis due to excessive usage of radio frequency spectrum is significantly emerging nowadays. And thus, a new technology is needed to exchange data and communication between devices, especially in smart homes and work regions. The expected method is invented by Harald Haas from The University of Edinburg, United Kingdom. He introduced this technology during his speech on TED Global in 2011. It is called as Li-Fi (Light Fidelity). Li-Fi basically depends on transmitting the data signals via light instead of radio waves. The data is represented as "1's" and "0's" in computer science. In this technology, if the LED is on, it will represent a 1, and if the LED is off, it will represent a 0. It has capability to transfer 4 Gigabit per second, and this opportunity is supplied by a single LED light. LED can also be used for illuminating the place during the data exchanging because on/off switching of LEDs cannot be noticed by human eyes. As well as the speed, it also supplies innocuous delivery unlike radio spectrums. Most of the emerging new smart buildings wont use Wi-Fi due to blocking monitoring equipment by radio waves. Alternatively, Li-Fi will be used. Since the light can be restrained easily, it gives a chance to protect the network from unwanted usage, and security problems.

WiMax- It stands for Worldwide Interoperability for Microwave Access, which is a broadband wireless access technology that supports fixed, portable, and mobile access. It can most simply be defined as the developed Wi-Fi. When it was produced first, the biggest property of Wimax was that it can broadcast to an area of 75 k/m with the fast of 30–40 Mbit/s to provide internet services. It means that by using a few Wimax antennas, people can have a high speed, wireless internet connection everywhere in the city and the smart homes region. Wimax was intended to provide a broadband connectivity to mobile devices. The mobile version of the technology was produced in 2005, known as IEEE 802.16e. It was embedded into tablet PCs, laptops, and smart phones. Accordingly, these devices are connected to internet while walking or setting in a moving car [4]. It would not be as fast as wired internet, but it was expectd to provide about 15 Mbps capacity in a 3 km cell coverage area. One of the Wimax advantages is that it has removed modem and cable requirements. With the developing technology, Wimax also has developed and wimax 2 (the IEEE 802.16) which has been produced in 2012. It eventually reaches the speed of 360 Mbit/s. However, the connection speed decreases according to users' count. If the channel used is intense, the adaptive modulation can keep the users connected. Also, wimax provides Low-cost and high-speed Internet access for individual users. In addition, Wimax provides safer internet Access. It has advanced error correction feature for safer communication.

IV. APPLICATIONS

Many applications that can be employed in smart homes are propose in [10]. The most common applications of home automation are lighting control, HVAC, outdoor lawn irrigation, kitchen appliances, and security systems. Occupancy sensor adjusts the temperature and turns off lights when a room is not in use. Window contacts setback HVAC when windows or balcony doors are left open. Wall switches are used to control lighting and shading. Heat valve is for self-powered and energy-efficient room temperature control. Room temperature sensor is used for minimal energy consumption and maximum comfort. Plug-in receiver controls and monitors consumer appliances. Motion activated sensors, as the name implies, these types of home security sensors are used to detect an intruder's presence. It is used for smart home protection. Once a sensor is tripped by a thief entering your home, a signal is sent to the smart home panel. In addition to setting off the internal siren, the smart home panel will also send you a text message or email, informing you of events as they unfold. Door contacts, once armed, a signal is sent to the smart home control panel if the door is opened.

V. CLOUD ARCHITECTURE FOR SMART HOMES

Cloud Computing is a recently existed technology trend with the aim to deliver computing facilities as Internet services. There are many thriving commercial Cloud services examples built by existing companies such as SaaS, PaaS, and IaaS, to merely mention a few. However, these services are all computer-based and mainly designed for the services provided by web browsers. Currently, we cannot find a cloud architecture that can provide the users with special services to the digital devices installed in smart homes. In this study, we propose an additional cloud Model, namely the smart home Cloud, which relies on the current cloud architecture with a modification of the service layer. This will result in providing efficient and stable services for the smart homes' owners. On the other hand, we recommend a Web service and Peer-to-Peer (P2P) technologies (see Figures 6.1 and 6.2) to the Cloud. This way, the cloud server will be able to provide a better service for the higher-quality audio/video signals by reducing the bandwidth pressure while transmitting them. In addition, the smart home gateway is responsible for both describing their services in WSDL and registering them the in cloud service directory. This way, the other homes can search for the service and benefit from it. Consequently, we can consider smart homes as service consumers and suppliers at the same time. Peer-to-Peer (P2P) and Web service are reasoned technologies that can be introduced to the cloud for combining both the cloud and smart homes. By using these two technologies, the cloud successfully provides a more special functionality to the smart homes. Recently, the home gateway has experienced too much

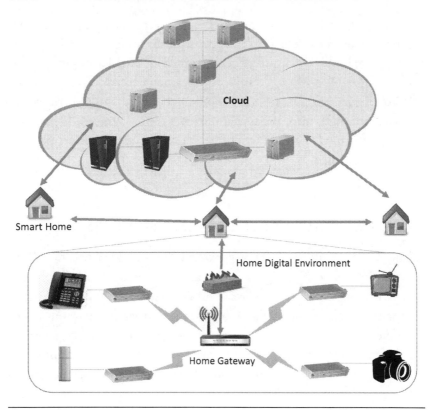

Figure 6.1   P2P network diagram.

research focusing on it. One of the initiatives resulted is the Open Services Gateway Initiative (OSGi). This initiative intends to create a platform that enables deploying services over both the wide-range network and local network or device. In the OSGI architecture-based smart home, the residential service gateway has two main functions, connecting the appliances inside the house, and linking the external network to the outside of the house. This way, People can experience a better home life without the need to overpower the smart devices with complex technologies and spontaneous user interfaces [14].

A. Peer-to-Peer Networks

In this part, we consider the smart home services as classified into three main categories, namely home entertainment, video communication

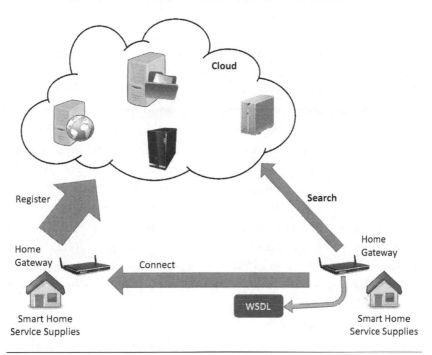

Figure 6.2 Web service architecture diagram.

and video conferencing, which are all based on audio/video streaming. The traditional client-server cloud model is constrained by the problem of bandwidth load pressure while transferring heavy high-quality audio/video signals. Solving such a problem required us to propose a mechanism where Cloud server and the smart home nodes can set a peer-to-peer network. This will reduce the bandwidth loads on the cloud. For this model, each user will be responsible to register some information to the cloud such as his name and IP. Moreover, a specific software for the P2P communication should be installed in the gateways of each home. As a result, the home nodes can share the bandwidth loads and the overall processing power. For instance, a user can set a public cloud to discern the real-time HD video broadcast for the other users who share the same cloud with him. Further, the other users can benefit from such a service by watching this video. So, the participant who is broadcasting the video stream is a customer downloading the video and a seller, who is uploading it to the others, at the same time.

B. Web Services

The traditional client-server model, the homes are considered as consumers of services only. In contrast, in our model, the smart homes (Peers) are consuming services and sharing them with the others. So, we can consider them as both consumers and suppliers of a service at the same time. To understand this feature, a Web service technology is introduced [14]. It uses Extensible Markup Language (XML) messages and follows the Simple Object Access Protocol (SOAP) standard. The home gateway is responsible for connecting and manages the various devices and networks installed within the house such as home automation network, and PC network, as a result, this can create an integrated digital environment. Moreover, this model enables the users to use the home gateway to share some features provided by their appliances with the neighboring houses. This can be achieved by describing the required services in a Web Services Description Language (WSDL) and distribute them into the cloud. Once the services are shared in the cloud, the other users can search for and use them. Hence, the cloud is considered as a UDDI server in the system, in which the service providers can publish their services and the customers to discover them. Further, it allows them to define how the services and the applications interact with each other on the Internet.

In Figure 6.2, the proposed Web service in the smart home architecture is shown. For instance, if a user has a printer in smart home A and he wants to share one of its services with the other houses, the house service gateway construes the service related to the online printing in WSDL and register it to the Service Directory in Cloud. Other smart houses can benefit from this service after paying to the owner of Home A. So, smart home A can make a profit from publishing his service on the cloud.

VI. ARCHITECTURE OF SMART-HOME–BASED CLOUD

The cloud architecture can be divided into three parts, namely the platform layer, the infrastructure layer, and the service layer as shown in Figure 6.3. The platform layer with infrastructure layer creates the base for delivering traditional IaaS and PaaS. On the other hand, the service layer is responsible for directly interacting with the smart home and focuses on the application services (SaaS).

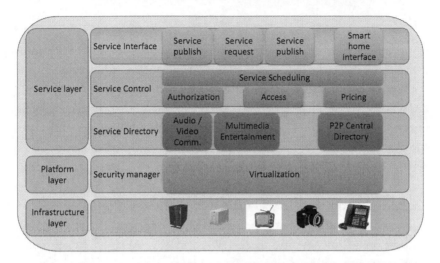

Figure 6.3 The three-layer cloud hierarchy for smart homes.

A. Service Layer

This part includes three main parts, namely Service interface, Service directory, and Service control. Service interface directly interacts with the smart home users. Furthermore, the users can easily publish their service description using the service interface. Service control part, which is considered as the brain of the service layer, is responsible for analyzing, processing and responding the user's requests.

B. Platform Layer

This layer is considered as the essence of the cloud model presented above. It includes two main modules, specifically security manage module, and resources management module. Further, Platform Layer with the help of Infrastructure Layer can create the base for delivering PaaS to smart homes.

C. Infrastructure Layer

The Infrastructure Layer contains a huge amount of physical resources responsible for delivering the cloud services. These resources are handled by a higher-level virtualization component, which is controlled by the cloud to provide IaaS Services for smart homes.

VII. PRIVACY & SECURITY ISSUES

To improve the user's privacy and comfort, the ULD system is a must and is designed to be more user-friendly than conventional localization approaches. Figure 6.4 shows a high-level architecture for supporting context-aware services in future smart homes. Also, it clearly presents the role of our ULD system in the architecture. Several types of sensors monitor the home environment to detect the user(s). Each sensor that detects a user sends a witness signal (as context information) through a sensor network via heterogeneous interfaces (e.g., ZigBee, Bluetooth, WLAN) to a Home Gateway (HGW).

Portable interchanges frameworks have developed through remote innovation advancement into 2G, 3G, and afterward 4G to keep pace with ever expanding voice and information activity. Harder security instruments are set up to shield today's portable correspondence frameworks. For example, one-path confirmation in 2G has been hoisted to shared validation in 3G and 4G; key length and calculations are getting to be more hearty; as versatility administration is enhancing, a forward key division in handovers has been included 4G; additionally, more successful security assurance is considered.

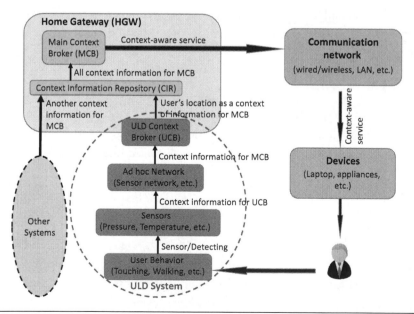

Figure 6.4   A high-level architecture for context-aware services in future smart homes [15].

Conventional security structures concentrate on insurance of voice and information, and they all have the accompanying security highlights in alike manner: 1) User identity management based on (U) SIM, 2) Mutual authentication between networks and users, and 3) Securing the path between communicating parties hop-by-hop.

Since the essential objective is to improve individuals' life through correspondence. Clients may impart by instant messages, voice calls, and video calls, or surf Internet or get to application administrations utilizing advanced cells. Notwithstanding, 5G, one of the key emerging smart homes technologies, is at no time in the future limited to individual clients. It's not just about having a quicker versatile system or wealthier capacities in advanced mobile phones. 5G will likewise serve vertical businesses, from which a differing qualities of new administrations will stem. With regards to vertical industry, security requests could change fundamentally among administrations. For example, portable Internet of Things (IoT) gadgets require lightweight security while fast versatile administration's request high productive portable security. The system based bounce by-jump security approach may not be sufficiently effective to fabricate separated end-to-end (E2E) security for various administrations. As IoT is picking up energy, more individuals will have the capacity to remotely work or "talk" to organized gadgets, for example, teaching offices at a savvy home to get up. Along these lines, there is a need of a more stringent verification strategy to anticipate unapproved access to IoT gadgets. For instance, biometric distinguishing proof could be a piece of the confirmation in savvy homes.

New IT advancements, similar to virtualization and Software Defined Network (SDN)/Network Functions Virtualization (NFV), are viewed as a way to make 5G arranges defter and more effective, yet less exorbitant. While CT are upbeat to see IT infusing new power into their systems, new security concerns are still developing. Security can't be worked for 5G administrations unless the system foundation is hearty. In heritage systems, security of capacity system components depends generally on how well their physical elements could be disconnected from each other. In 5G, the seclusion will work contrastingly as virtual Network Elements (NEs) on cloud-based framework. It's imaginable that time is all in all correct to think about 5G framework security. SDN is turned out to be of assistance in enhancing transmission proficiency and asset design. Then again, it is imperative to consider in the 5G security outline that it could be overseen as far as the seclusion for system hubs, for example, control hubs and sending hubs, and the protected and right authorization of the SDN stream table. In light of system virtualization innovation, a system could assemble diverse virtual system cuts. Each virtual system cut could oblige a specific administration prerequisite and in this way may require separated security capacities. 5G security configuration may thus require to consider issues of how to segregate, convey, and oversee virtual system cuts safely.

The heterogeneous nature of the communication paradigm in the emerging smart homes comes not just from the utilization of various wireless technologies, additionally from multi-organize conditions. IoT gadgets have numerous options in the way they reach to the smart home systems. For example, they may interface with systems straightforwardly, or through a portal, as in the D2D or the Relays form. Contrasting with portable handset, security administration of IoT gadget in 5G might be productive and lightweight so as to build up trusted connections amongst the typical gadgets and smart homes systems. However, 5G systems raise genuine worries on security spillage. As a rule, protection spillage can bring about genuine results. As essential strategy for system getting to, versatile systems conveys information and flagging that contains numerous individual protection data (for example, character, position, and private substance). With a specific end goal to offer separated nature of administration, systems may need to detect what type of administration a client is utilizing. The administration type detection may include client security.

Meanwhile, Li-Fi provides better security against the Wi-Fi. The Li-Fi signal can be blocked by the walls or any other opaque objects. Therefore, the coverage can be controlled, and unauthorized accesses can be prevented. The motto is "If you cannot see the light, you cannot access the data". This increases the security, but also affects the mobility. Li-Fi can be applicable inside a room, but there will be a problem for using it outside a building. The best solution would be in this case to use Li-Fi repeaters (like Wi-Fi repeaters).

On the other hand, WiMax uses security certificates which make it difficult for an attacker to spoof the identity of a legitimate subscriber, providing ample protection against theft of services. Nevertheless, like the 802.11, management frames are not encrypted, allowing an attacker to collect information about subscribers in the area and other potentially sensitive network characteristics. Despite the good intention for Wimax security, there are several potential attacks open to adversaries, including Rogue Base Stations, and Denial of Service (DoS) Attacks.

In general, IoT products include poor encryption and back doors that could allow unauthorized access. The cost of manufacturing some of the sensors and other devices constructing the IoT paradigm imposes some restrictions in terms of the ability to update/improve these devices. In addition, visibility of such devices could be close to none. This creates a security vulnerability when a user believes an IoT device is performing certain functions, when in reality it might be performing unwanted functions or collecting more data than the user intends. The challenge here is thus the privacy of data collected by IoT devices. Where the clients agreeing to data being collected is not as straightforward as it was previously. The nature and frequency of data collected or sent by the thing makes it harder for the client to get involved. Second, some IoT devices do not commend an interface to configure privacy settings. The data collection and processing itself can always affect people and devices not involved with the IoT ownership. Examples are people sharing the same geographic location temporarily or sharing the same computing resources. Additional privacy protection measures are thus required for every carrier of a smart device because as the time goes on there is more of an assumption that each piece of data that can be collected will be collected by another device around you.

VIII. CONCLUSION

Smart Home applications provide its homeowner's comfort, security, energy efficiency (low operating costs) and convenience for the users. The smart home system offers solutions to problems such as; fire, flood, gas leak, theft, child/elderly care, smart door, temperature and humidity control, security management, air conditioning and irrigation control, curtain-shutter control, light controller etc. Any devices can work in harmony with each other to provide a better life quality. In addition to more security, the smart home system allows you to administer electrical devices such as TV etc. remotely. Moreover, it provides you with reports to prevent further energy consumption. You can control all these high security and comfort services from wherever you have Internet access with a smartphone, tablet or computer. For the user, these services will be time efficient and more secure, because human cannot remember everything. Therefore, the automated system can control all housework in an instant of time. For example, in terms of security management, smart homes are very useful. Thanks to smart home applications, users can do some regulations for security management. Therefore, it will be by remote control. It will be cost-efficient and safer.

REFERENCES

[1] D. J. Cook, A. S. Crandall, B. L. Thomas, and N. C. Krishnan, "CASAS: A smart home in a box," Computer, vol. 46, no. 7, pp. 62–69, 2013.

[2] A. Al-Fuqaha, M. Guizani, M. Mohammadi, M. Aledhari, and M. Ayyash, "Internet of things: A survey on enabling technologies, protocols, and applications," IEEE Communications Surveys & Tutorials, vol. 17, no. 4, pp. 2347–2376, 2015.

[3] R. Want, B. N. Schilit, and S. Jenson, "Enabling the internet of things," Computer, vol. 48, no. 1, pp. 28–35, 2015.

[4] D. Yang, B. Xu, K. Rao, and W. Sheng, "Passive Infrared (PIR)-Based Indoor Position Tracking for Smart Homes Using Accessibility Maps and A-Star Algorithm," Sensors, vol. 18, no. 2, p. 332, 2018.

[5] D. Sasakawa, N. Honma, T. Nakayama and S. Iizuka, "Human Identification Using MIMO Array," in IEEE Sensors Journal, vol. PP, no. 99, pp. 1–1. doi: 10.1109/JSEN.2018.2803157.

[6] https://www.electronichouse.com/smart-home/home-automation-protocols-what-technology-is-right-for-you/

[7] Brian Ray, "Applications of Home Automation", Published March 17 2015. https://www.linklabs.com/blog/applications-of-home-automation.

[8] M. B. Yassein, W. Mardini, and A. Khalil, "Smart homes automation using Z-wave protocol," in: IEEE International Conference on Engineering & MIS (ICEMIS), IEEE, 2016, pp. 1–6.

[9] M. Siekkinen, M. Hiienkari, J. K. Nurminen, and J. Nieminen, "How low energy is bluetooth low energy? comparative measurements with zigbee/802.15. 4," in: IEEE Wireless Communications and Networking Conference Workshops (WCNCW), IEEE, 2012, pp. 232–237.

[10] P. Gaikwad, J. P. Gabhane, and S. S. Golait, "A survey based on smart homes system using Internet-of-things," in: IEEE International Conference on Computation of Power, Energy Information and Commuincation (ICCPEIC), IEEE, 2015, pp. 0330–0335.

[11] D. Valtchev, I. Frankov, "Service gateway architecture for a smart home", *IEEE Communications Magazine*, pp. 126–132, 2002.

[12] Dong-Sung Kim, Jae-Min Lee, Wook Hyun Kwon, "Design and Implementation of Home Network Systems Using UPnP Middleware for Networked Appliances", IEEE Transactions on Consumer Electronics, vol. 48, no. 4, November 2002.

[13] T. Gu, H.K. Pung, "Toward an OSGi-based infrastructure for context-aware applications", IEEE Pervasive Computing, pp. 66–74, 2004.

[14] E. Newcomer, Understanding Web Services: XML Wsdl Soap and UDDI, Addison-Wesley Professional, 2002.

[15] Ahvar, E., Lee, G. M., Han, S. N., Crespi, N., & Khan, I. (2016). Sensor network-based and user-friendly user location discovery for future smart homes. Sensors, 16(7), 969.

7

IoT-BASED EDGE AND SECURITY: DISCUSSIONS AND REMARKS[1]

Contents

---

[1] Previously published in F. Al-Turjman, "Cognitive Caching for the Future Sensors in Fog Networking", *Elsevier Pervasive and Mobile Computing*, vol. 42, pp. 317–334, 2017.

139

Abstract

To the best of our knowledge, there is not enough substantial investigations on security issues in the IoT-based Edge Computing. Therefore, in this chapter, the impact of Edge computing in IoT environments is analyzed from security and privacy perspective. We aim at proposing a new direction of the IoT research on threats and security attacks detection and prevention. Furthermore, formal fidelity and trust analysis will be discussed as well to emphasize the effectiveness of Edge computing in IoT paradigms, where edge devices are prone to unsecured data retrieval from the authorized nodes in the cloud.

Keywords

Internet of Things, Edge computing, Caching, Security, Cyber Attack.

I. INTRODUCTION

Internet of Things (IoT) has paved ways for connecting devices, people, infrastructure, and practically any entity that can be connected. Cloud infrastructure has been the irrefutable part of IoT solutions with substantiated claims of resource provisioning, scalable platforms, and minimal infrastructure costs. Cloud computing does not argue well with applications that require reduced latency, positioning capabilities, and support for mobility. Thus, a new platform is needed to meet these requirements. A new platform, called Edge Computing [5], or, simply, Edge, because the Edge paradigm is a cloud close to the edge of the Internet proposed to address the aforementioned requirements. Edge is a Mobile Computing (MEC) platform that puts services and resources of the cloud closer to users to be facilitated in the IoT networks.

Unlike Cloud Computing, Edge Computing enables a new breed of light applications and services, that can be run at particular terminal networks, such as WSNs and RFIDs. In order to enable WSNs to support this trend in communication and function in a large-scale application platform, such as the *Edge Computing*, we proposed the cognitive framework in our previous work [7]. In [7], an information-centric scheme is proposed for the future WSNs using *cognitive* in-network devices that makes dynamic routing decisions based on specific *Knowledge-* and *Reasoning-* observations in WSNs. AHP is applied on quality of information (QoI) attributes in next generation WSNs such as reliability, delay, and network throughput observed over the communication links/paths [15][16]. This cognitive Information-Centric Sensor Network (ICSN) framework is able to significantly outperform the *non-cognitive* ICSN paradigms. However, this cognitive ICSN framework did not consider yet the in-network caching feature. Caching in multitude of nodes in the Edge paradigm has pivotal role in enhancing the IoT network performance in terms of reliability and response time. In this chapter, we discuss possible caching techniques in edge-based IoT networks in addition its security issues and influences.

The cache replacement strategies in the literature [17][18], have been strategically designed so far for IP-based networks, which have significant variations against the targeted vision of the future Edge networks. Where Edge is emerging as one of the most promising content-oriented networks that necessitates massive changes in the core architecture of the system. Foremost thoughts must be specifically put into the cache replacement strategy, as it counts to dramatic influence on the network performance. To this end, a novel cognitive caching framework which can serve several kinds of applications in the cloud with content-oriented requirements is needed.

The remainder of this paper is organized as follows. Section II provides a literature overview on the Edge caching techniques and categories. Section III, key design factors in IoT-based edge computing are discussed. In Section IV, we introduce our IoT-based Edge framework. Section V presents a formal prove about the proposed framework security in edge computing. Finally, Section VI concludes our work.

II. CACHING IN IoT-BASED EDGE

At the core of the Edge paradigm, which comes with plenty of heterogeneous edge devices, data has to be located close to the device requesting it. This is the mission of the caching scheme usually running over the cloud nodes. Since the environment, the type and the amount of data received varies from one device to another over the cloud, it is hard to come up with a single caching strategy that fits everyone's needs. Therefore, there are several caching strategies that have been proposed under various assumptions for different scenarios. However, efficient caching strategies most often demands two main properties, first, an updated copy of the requested data must be always close to the area of interest. Second, the data copy shall stay "alive" as long as it might still be used. IoT network caching in the literature can be categorized into the following: *A) Functionality-based caching, B) Content-based caching, and C) Location-based caching.*

A. Node Functionality-Based Caching (FC)

To maximize the full potential of the content oriented networks, we must consider which content should be stored in the control level rather than guessing it at the data level. Consequently, the authors in [34] claim that there are side-effects for handing over the caching decision to the data level, and they propose an approach to handle the data at the control level. However, their approach is not applicable in large-scale scenarios. Authors in [35] propose an information-centric caching algorithm known as LocalGreedy, where a cluster of caches is considered, with different leafs either connected directly or indirectly via a parent node. An inter-level cache cooperation is utilized in order to fetch the data from a specific node called the parent node and not from any other node. This approach employs a strong conceptual similarity check to provide more cost-effective solutions especially when the idea of 'access cost' is adopted. The access cost is represented either by the accrued cost of latency when fetching data from remote caches, or by the consumed bandwidth when retrieving content from peer nodes. However, this method necessitates the knowledge of the in-network nodes' capabilities and this contradicts with the Fog vision. Authors in [36] have studied the trade-off between caching the content in distributed IP-based networks and the new emerging content-oriented architectures in Fog systems. The sturdy is applied to a mixture of real traffic from sources like the web, file sharing and multimedia streaming, and it concludes that caching real-time contents in routers can increase the cache hits. Nevertheless, this caching system is not sufficient for Fog systems, where other content types would likely be more efficiently handled at more capable devices such as smartphones/laptops at the edge of the network.

B. Location-Based Caching (LC)

In order to achieve better network performance, authors in [19] recommends caching less, and argues against caching everywhere in ICSNs. This caching policy states that, data should only be stored in a node with the highest probability of getting a cache-hit. Cache Aware Target idenTification (CATT) is a proposed caching policy in [20] where a node is selected for caching as long as the node has the highest connectivity-degree based on its geographical location. However, this makes the node a geographical bottleneck in the network. Another location-based caching has been proposed in [21], where a probabilistic caching method combined with geo-factors is utilized. The proposed method is called pCASTING and considers a probabilistic least recently used (LRU) approach, as well as data freshness, device energy and storage capability. Authors show how these metrics are employed in a utility function to decide the caching option. They compare their results with no caching, with 50% probability to cache, and with the everything and everywhere caching strategy. As a result, they conclude that there is at least 10% difference in favor of pCASTING for the cache hit ratio, and a higher value of successfully received packets as well as lower delay.

Moreover, a topology-based replacement approach is proposed in [22] where replicas are placed at the intermediate routers of the Internet. Authors have found out that the router fan-out based replacements need to be carefully designed to maximize the cache efficiency in content-oriented systems, where self-organized caches are required. Self-organized caches are caches which make consistent decision about caching. Authors in [23] have considered this strategy and concluded that it is has better average delay compared with traditional methods (Cache at intermediate nodes), using smaller per-node caches.

Meanwhile, authors in [24] have proposed another caching method, namely the selective neighbor approach, which selects the most appropriate neighboring proxies for minimum mobility overhead in terms of average delay and caching cost. This approach is based on proactively caching data requests and the corresponding meta-data to a subject of proxies that are one hop away from the proxy. Authors in [26] suggests a probabilistic approach for ICNs. They claim that the probability of a file being cached should be increased as it travels from source to destination by considering the following parameters: i) The distance between source and current node, ii) Distance between destination and current node, iii) Time-to-Live for the routed data content, and iv) the Time-Since-Birth. Authors also suggests redundancy in caching on a single path between source and distention. However, this degrades the ICN performance dramatically while experiencing limited caching spaces. Moreover, in [26], authors assume that all the network nodes has the capability of caching, which is not the case in practice with Fog systems. The proposed approach is weak as well due to considering static data request's frequency from a subnet where that data can exist. Nevertheless, we believe caching should be based on dynamic frequencies and location-independent.

C. Content-Based Caching (CC)

Content-based caching is another category for caching in content-oriented fog networks, in which the data replacement decision is taken based on the content of the exchanged data. For example, an automatic cache management system that dynamically assigns data to distributed caches over the network has been proposed in [27]. In this reference, distributed file managers make the data replacement decision based on the observed request patterns. This approach assumes that every cache manager has access to all the caches and data request patterns. Consequently, this approach experience reduction in access time at the cost of additional massage exchange and computational overhead which can dramatically degrade the fog performance.

Another approach, proposed by Hail et al. (2015), suggests that LRU would most likely be the best candidate strategy for caching to be used in a heterogeneous information-centric network such as the cloud [29]. Authors in [29] compare the pure LRU strategy with three other strategies; the pure randomness, the probabilistic LRU and the probabilistic caching method pCASTING. According to their results, the probabilistic LRU performs better than the other methods in terms of energy, and environmental metrics.

Authors in [30] aims at minimizing the data publisher load and maximizing the in-network cache hit by storing frequently requested data on selected routers over the network. The authors have presented two popularity-based caching algorithms from the basis of optimal replica replacement. However, this work may not be practical as authors have only considered one gateway in the network, while plenty of them can be utilized in Fog paradigms. WAVE is another caching approach in which the cache size is adjusted according to data popularity features [31]. In this approach, a master upstream node recommends the count of data chunks to be cached at the downstream slave nodes. This count increases exponentially with the number of requests in order to avoid any unnecessary overhead. WAVE distributes data over the network edge while considering popularity to distance relationships. However, the different volumes of data chunks have not been deliberated in this work.

Authors in [32] propose an age-based caching approach, which aims at reducing data publisher load and the in-network delays. This approach provides the techniques where ages of the contents/data are dynamically updated. It distributes popular content to the edge of the network while eliminating unnecessary replicas at the intermediate routing nodes. However, this approach fails to handle frequently changing contents, and hence, in-network devices which are far away from the data publisher may experience extended delays.

Accordingly, the design of the cache replacement policy must be a dynamic one based on the user request pattern and the implemented application. In the IoT-based Edge paradigm, typical caching strategies cannot be applied without extra manipulations. That is simply because of the security issues, the uncertainty of the wireless medium, and the disparate need for user requirements-awareness in Edge-specific IoT. Privacy, storage capacity, and heterogeneity in terms of sensors and other utilized enabling technologies at the edge devices are further challenges to be considered. Additionally, there are some factors that the replacement policy should take into consideration.

III. DESIGN FACTORS

In this section, we list key design factors and abridge our IoT-based Edge vision accordingly.

A. Latency in Data Access

Different sensors and edge devices experience varying latencies to be exposed to the environment and to effectively send back their data. Accordingly, we have to store data for extended periods of time when the experienced delay in gathering data from the surrounding environment is predicted to be increasing not decreasing. This is called the sensing delay. Additionally, propagation delay would be added if every time that data is requested, it has to be moved from SNs/edge nodes to an IoT user, especially if the edge nodes are located far away from the Sink/data publishers in the Cloud. The propagation delay τ a packet encounters is generally proportional to the number of hops between the packet's source and destination. The number of hops between these two points can be approximated to be a linear function of the Euclidean distance between them. Therefore, the propagation delay a data packet encounters can be expressed by

$$\tau = \alpha \, L \tag{7.1}$$

where α is a constant and L is the Euclidean distance between the packet's source and destination. With a stationary data requester located at the center of the edge of the cloud, the maximum propagation delay a packet may encounter occurs when the source is at the center of the cloud; this results in a delay of αR, where R is the radius of the user cloud. On the other hand, the maximum delay in a network with a mobile data requester is significantly worse. With a mobile data requester, the worst case occurs when a packet arrives to a relaying node which has just been left by the requester/user; such a packet needs to wait for the data requester to complete a full round along the perimeter of the cloud which depends on the speed of the data requester and on other factors. It is obvious that such a delay is much longer than that of a network with a stationary data requester/user. The total delay (Δ), which involves delivering fresh data to the user is a combination of the sensing and propagation delay, given by Equation (7.2) below.

$$\Delta = \tau + \delta \qquad (7.2)$$

B. Data Age Factor

The recommended age factor in an IoT-based Edge paradigm makes use of the following two conditions to decide which content shall be removed from the edge device cache memory. The first is based on the validity duration of the periodic request (Type I traffic type in Table 7.1), and second, when the edge device cache is full. We make use of this duration interval, because freshly retrieved data has to be provided at the start of each periodic request cycle [28]. Thus, when the cache is full at the end of each periodic request cycle, old data can be replaced with more recent ones. Thus the age attribute-value pair is represented by its time-to-live (TTL) which is based on the periodicity of the request for each application type. In the following, Equation (7.3) represents the TTL_c of the cached data (c) represented

Table 7.1 QoI Priority for Different Data Traffic Types

REQUEST TYPE	LATENCY (L)	ENERGY (E)	RELIABILITY (R)	THROUGHPUT (T)
Type I: On-Demand	x	3	1	2
Type II:Periodic	1	2	4	3
Type III:Emergency	1	1	x	2

as attribute-value pair, is directly dependent on the periodicity of a request in Type I traffic flow.

$$TTL_c \propto T_{periodic} \tag{7.3}$$

In case the application requires that the periodic data is stored for a prolonged duration of time, for example 24 hours, before making a single transmission to the sink, then the cache retention period becomes a function of the duty cycle's periodicity.

C. Popularity of the Data Request

Traffic flow generated in response to on-demand requests have been classified as Type 2 traffic. More number of users may be interested in a particular type of sensed data, or a specific sensed data may be requested more number of times by one or more users. Such sensor data is said to be popular, and can be retained for prolonged time periods in the edge device's cache. Thus the popularity of the cached data attribute-value pair can be represented by Equation (7.4) below.

$$\text{Popularity}_c \propto \text{Req}_c / \text{Req}_{total} \tag{7.4}$$

where Req_c is the total number of requests for an attribute-value pair received at an Edge device, and Req_{total} is the total number of requests received by that device, within a particular periodicity cycle. In addition, when the in-network devices' batteries start to be depleted, Edge devices should store the data for prolonged time periods to maintain their availability. When the edge device storing such data starts to die out, neighboring devices can be good alternatives to store the data and/or provide extra storage for guaranteed availability.

D. Trust Factor

Since we are relying heavily on distributed resource which can be owned by different parties in the Edge paradigm, it is necessary to verify the fidelity of the retrieved data in order to avoid any unsecure/improper access. And thus, a Trust factor (T) must be introduced in this domain. This factor is calculated at a gateway per user to represent a GW_j fulfillment measure. A higher T_{GW_j} indicates that previous data exchanges between $user_j$ and GW_j have been fulfilled according to attributes promised by GW_j. We hence mark some data to be trusted while other data may not. The calculation of T_{GW_j} could follow a function similar to the fuzzy reputation formula presented in [32]. Accordingly, we can assume an arbitrary value between 0 and 1 to express the trust parameter according to the following function:

$$T = \left(T_{GW_i}\right)^{\gamma} \tag{7.5}$$

where $\gamma > 0$ is a weight variable to control the slope of the function by giving more emphasis to the trust parameter.

E. False Data Injection Factor

For any IoT-based Edge system, data security and data aggregation is a vital issue. The Edge system performance, effectiveness and efficiency all are related to data aggregation and acceptance in related to data privacy. These are one of the vital issues for a cyber-Edge system and also become the main target of cyber-attacks. With the evolution of the technology, cyber systems are upgrading and new types of cyber-attacks are discovered too. However, a particular type of cyber-attacks, known as False Data Injection Attack (FDIA) is the riskiest one in IoT-based Edge paradigms. False Data injection is a cyber-attack where the compromised host construct events which do not take place in that instance of time. In this type of attacks, the attacker can take advantage of the small error rate tolerated by the system algorithms and gradually increase the impact of the injection so that the increase of fake data is undetectable. FDIA require strong analysis of the targeted system by the attacker who should know the topology of the IoT network. Another FDIA requirement is that the attacker should have physical access to temper in the system. Accordingly, FDIA has major impacts on the system. It is a one of the major cyber-attacks which can cause high level damage by altering the data aggregation and creating false results.

IV. IoT-BASED EDGE FRAMEWORK

In this section, we detail and discuss our visional architecture for the intended future IoT-based Edge paradigms. Figure 7.1 shows the

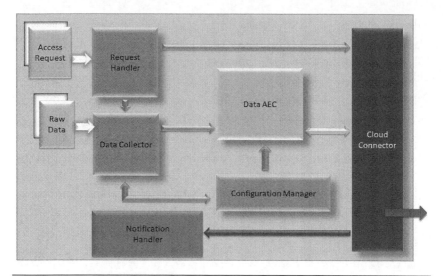

Figure 7.1 Edge Device Architecture.

client-based edge device modules: 1) the input and output proxies, 2) the data collector, 3) the data aggregator, 4) configuration manager, 5) request handler, 6) notification handler and 7) the cloud connector, described as follows:

1. Input Proxy: As the IoT Edge applications can be installed on different devices (e.g., smart phones, Conitiki devices, and CoAP-based devices) that differ in their capabilities and adopted communication protocols, input proxy module is used to handle any incoming communication with the device module. This is needed to isolate the Edge device from the used intermediate devices' technology and handling details.

2. Request Handler: It is the module responsible of getting the requests from the inhabitants to access any other devices or services. The request handler prioritizes the concurrent requests if any and send them to the cloud.

3. Data collector: It is the module responsible of monitoring all the inhabitant's interactions on the smart device, and collecting the corresponding raw data, then sends it to the aggregator module.

4. Data Aggregator: It is the module responsible of aggregating the raw data of the interactions into blocks, then it should

encrypt them (if needed), then compresses such blocks to minimize the transfer time between the smart device and the cloud. The output of the data aggregator module is blocks of Aggregated, Encrypted, and Compressed (AEC) raw data. Accordingly, it allows the encryption and compression operations to be optional at the client side, as not all smart devices can handle such operations.

5. Cloud Connector: It is the module responsible for sending the AEC raw data blocks to the cloud. So it should synchronize the transferred blocks with the communication proxy located in the cloud.

6. Configuration Manager: It is the module responsible for data configuration, in which the inhabitant specifies which interactions and information to be monitored and collected by the Edge device, also inhabitant can specify the preferred sampling rates, aggregation cache size, encryption and compression options. This module is intended to provide the inhabitants with the ability to control the privacy of its data, as they can specify which data to be encrypted and which data remains open for public, also the module provides inhabitants with the ability to control the power and storage consumptions on the Edge device.

7. Notification Handler: This module is responsible of collecting the notifications to be sent to the inhabitants, also it prioritizes these notifications coming from the cloud, and send them to the output proxy.

8. Output Proxy: This module is responsible to send the notification and warning messages to the inhabitant based on the used device capabilities.

The key attributes whenever we are trying to implement security solutions are authentication, integrity, confidentiality, trust, access control, data security, and privacy. The configurations performed by the network administrator and the network management information must be secured and isolated from the normal data flow. This is required since the edge devices are distributed across the network and the cost of maintenance is high. However, a new network concept called software-defined networking (SDN) can come to the rescue. The benefits of SDN include but not limited to reducing the burden of management and implementation, improving the scalability of the network, decreasing the cost of the Edge system maintenance, and enhancing the access control of network resources.

Like cloud computing, the data storage in edge platform is also outsourced. It is tough to ensure data integrity since there are a lot of possibilities for data loss and data modifications. If the data present in one device is compromised, the attackers could easily abuse the data stored to fulfill their own needs. Hence, provision for data storage auditing should be done.

The data computation performed at the edge servers and devices should be secured and verifiable. The security of computations can be ensured by using data encryption techniques, which prevents data visibility to any hackers/attackers. Since the microdata centers have the provisions to off-load some of the computations to other data centers, a mechanism to verify the computation results and establish trust between the two entities becomes a necessity. This enables users to securely search for user data amidst encrypted data. This helps in maintaining the secrecy of encryption.

Since the end devices are typically less powerful and have limited access to the surroundings, they can be easily tampered with. Any hacker may try to take control of a device and make it a rogue node to retrieve essential network management data. They can hinder the normal behavior by corrupting the device data and increase the frequency of data access by sending fake information. Hence, proper efforts should be made for assuring the security of the Edge device.

Enforcement of access control mechanisms can provide dual benefits of security and privacy. Since the edge platform is distributed and decentralized in nature, a good access control policy acts as a defensive shield to mitigate unauthorized device and service access. Access control helps in realizing the interoperability and collaboration among microdata centers that are provided by different service providers and are separated across different geo-locations. A robust access control mechanism is required to meet the design goals, accommodate mobility, low latency, and interoperability.

Intrusion detection techniques help in identifying malicious data entries and detect device anomalies. They can be used to carefully investigate and analyze the behavior of devices in the network and provide methods to perform packet inspection, which helps in early detection of denial-of-service attacks, integrity attacks, and data flooding, among others. The primary challenge when implementing intrusion detection for edge platform is to accommodate for scalability, mobility, and low-latency requirements.

V. FORMAL SECURITY ANALYSIS

This section is composed of stringent formal edge caching analysis of the proposed framework. The analysis shows that the proposed framework not only offers data availability and accessibility, but also prevents the various potential data loss, and unsecure access. Our IoT-based Edge framework offers a secure agreement between the customer C_i, and a relay/gateway LCN_i and it is proven using the hierarchical Kura architecture and BAN logic proposed in [10], [11], and [12], respectively. Accordingly, let us assume that X and Y be the requesting devices of data, P and Q be the data to be cached and s_k be the secret key while considering other notations as summarized in Table 7.2.

Based on [12], BAN logic assumptions/procedures can be formulated as follows:

Procedure 1: If X trusts that s_k is shared between X and Y and observes P. encrypted with s_k, then X trusts Y as a legal device to cache the requested data. This can be represented by the following formulas:

$$\frac{X \mid\equiv X \overset{s_k}{\leftrightarrow} Y, X \triangleleft \{P\} s_k}{X \mid\equiv Y \mid \sim P} \text{ and } \frac{X \mid\equiv X \overset{Q}{\leftrightarrow} Y, X \triangleleft \{P\}_Q}{X \mid\equiv Y \mid \sim Q} \quad (7.6)$$

Procedure 2: If only Y observes P, then X trusts that Y is sure of P as shown in Equation (7.7).

$$\frac{X \mid\equiv\neq P, X \mid\equiv Y \mid \sim P}{X \mid\equiv Y \mid\equiv P} \text{ and } \frac{X \mid\equiv\neq Q, X \mid\equiv Y \mid \sim Q}{X \mid\equiv Y \mid\equiv Q} \quad (7.7)$$

Table 7.2 Notation Used in Our Analysis

NOTATION	DESCRIPTION
$X \models P$	X trusts P
$\neq P$	P fidelity is assured
$X \models\Rightarrow P$	X takes the authority over P
$X \triangleleft P$	X recognizes P
$X \models\sim P$	P formerly trusted as X
(P, Q)	P or Q is an individual part of (P, Q)
$\{P\}_{s_k}$	P is encrypted using s_k
$\langle P \rangle_{s_k}^{Q}$	P is cached at Q
$X \overset{s_k}{\leftrightarrow} Y$	X and Y uses s_k to establish a link. Besides, s_k is totally secure; and thus can not be discovered by any in-network device except X and Y.
U_{ser_i}	i^{th} user
PID_j	Identity of jth user
SK_j	Secret key of jth user
LCN_i	i^{th} local cognitive node
CID_i	Unique identity of ith LCN
$H_1 : \{0,1\}^*$	Map to point hashing functional operation
$H_2 : \{0,1\}^*$	Secure collision free one way cryptography hashing function
x	Secret random integer controlled by GC_S
$E_{S_k}(.)$	Symmetric key encryption function
ΔTS	Expected delay transmission time
TS_S	Timestamp
$\|$	Concatenation operator
$\oplus$	Bitwise X-OR operator
S, r, y, z	Random integers $\in Z_q^*$
S_{k1}, S_{k2}	Secure session key
q	prime order integers
H_1, H_2	Hashing operators.
s_k	Secret key
GC_S	Global cognitive sink
RN	Relay node

Procedure 3: If X trusts P and Q, then X can cache P and Q. And thus,

$$\frac{X \models P \quad X \models Q}{X \models (P, \ Q)} \tag{7.8}$$

Procedure 4: If X trusts the key of P, then Y trusts (P, Q) as follows:

$$\frac{X \models\neq P}{X \models\neq (P, \ Q)} \tag{7.9}$$

Procedure 5: If X believes that Y can affect the cached data P and X trust Y in terms of P validity, then X trusts P. And this can be formulated as follows:

$$\frac{X|\equiv Y \Rightarrow P \quad X|\equiv Y|\equiv P}{X|\equiv (P, Q)} \tag{7.10}$$

In order to satisfy the security factor in Fog computing, the proposed CCFF framework must be able to meet the following objectives:

$$\textbf{\textit{Objective}}_1 : U_{ser_i} |\equiv GC_S| \equiv LCN_i \overset{sk}{\leftrightarrow} RN$$

$$\textbf{\textit{Objective}}_2 : U_{ser_i} |\equiv LCN_i \overset{sk}{\leftrightarrow} S_C$$

$$\textbf{\textit{Objective}}_3 : RN |\equiv U_{ser_i}| \equiv LCN_i \overset{sk}{\leftrightarrow} RN$$

$$\textbf{\textit{Objective}}_4 : RN |\equiv LCN_i \overset{sk}{\leftrightarrow} RN$$

Where, the structural flow of BAN logic is as follows:
1. Messages to be exchanged/cached:

$$M_1 : U_{ser_i} \rightarrow GC_S : \langle H_2(x \oplus SK_j)\rangle, \langle PID_j, H_2(x \oplus SK_j)\rangle_{x \in Z_q^*}$$

$$M_2 : GC_S \rightarrow RN : \langle Certify_j = S.H_1(PID_j \| H_2(x \oplus SK_j));$$
$$TS_j = H_2(PID_j \| y); H_j = H_2(TS_j);$$
$$V_j = TS_j \oplus H_2(x \oplus SK_j)\rangle_{x \in Z_q^*}$$

$$M_3 : U_{ser_i} \rightarrow RN : \langle Certify_j, V_j, H_j, x\rangle_{x \in Z_q^*},$$
$$where \ V_j = TS_j \oplus H_2(x \oplus SK_j) \ and$$
$$H_j = H_2(TS_j).$$

2. Transmitted/cached messages in cognitive form:

$$T_{M1} : U_{ser_i} \rightarrow GC_S : \langle PID_j, H_2(x \oplus SK_j)\rangle_{U_{ser_i} \overset{PID_j}{\rightarrow} GC_S}$$

$$T_{M2} : GC_S \rightarrow RN : \langle Certify_j, V_j, H_j \rangle_{U_{ser_i} \xrightarrow{PID_j} GC_S}$$

$$T_{M3} : U_{ser_i} \rightarrow RN : \langle Certify_j, V_j, H_j, x \rangle_{U_{ser_i} \xrightarrow{PID_j} GC_S}$$

3. Cached data in Hypotheses form:

$$H_{M1} : U_{ser_i} \vDash \neq (CID_i), \ LCN_i \vDash \neq (TS_1, TS_2) H_{M2}$$

$$: GC_S \vDash \neq (PID_i), \ GC_S \vDash \neq (TS_3, TS_4)$$

$$H_{M3} : U_{ser_i} \vDash GC_S \vDash LCN_i \xleftrightarrow{sk} RN$$

$$H_{M4} : U_{ser_i} \vDash LCN_i \xleftrightarrow{sk} RN$$

$$H_{M5} : RN \vDash U_{ser_i} \vDash LCN_i \xleftrightarrow{sk} RN$$

$$H_{M6} : RN \vDash LCN_i \xleftrightarrow{sk} RN$$

Consequently, the proposed cognitive CCFF framework's security and fidelity can be examined and proved based on BAN logic assumptions' and objectives' as follows:

Proof. Based on the transmitted/cached message T_{M1}, the CCFF has $P_1 : \langle GC_S \vartriangleleft PID_j, H_2(x \oplus SK_j) \rangle_{U_{ser_i} \xrightarrow{PID_j} GC_S}$. And from H_{M2}, P_1 and Equation (7.6), the CCFF acquires $P_2 : GC_S \vDash U_{ser_i} \vert \sim \langle PID_j, H_2(x \oplus SK_j) \rangle$. And since the transmitted/cached message T_{M2}, the CCFF has $P_3 : RN \vartriangleleft \langle Certify_j, V_j, H_j \rangle_{U_{ser_i} \xrightarrow{PID_j} GC_S}$. Now based on H_{M5}, P_3 and Equation (7.6), the CCFF acquires $P_4 : LCN_i \vDash RN \vert \sim \langle Certify_j, V_j, H_j \rangle$. And from H_{M1}, P_4, Equation (7.7) and Equation (7.9), the CCFF obtains $P_5 : U_{ser_i} \vDash GC_S \vDash LCN_i \xleftrightarrow{sk} RN$. This validates our target in **Ojective₁**. And based on H_{M5}, P_5 and Equation (7.6), the CCFF gets $P_6 : U_{ser_i} \vDash LCN_i \xleftrightarrow{sk} RN$. This validates our target in **Ojective₂**. Now from the transmitted/cached message T_{M3}, the CCFF has $P_7 : GC_S \vartriangleleft \langle Certify_j, V_j, H_j, x \rangle_{U_{ser_i} \xrightarrow{PID_j} GC_S}$.

And from H_{M1}, P_7 and Equation (7.6), the CCFF acquires $P_8 : GC_S | \equiv U_{ser_i} | \sim \langle Certify_j, V_j, H_j, x \rangle$. From H_{M5}, P_8, Equation (7.7) and Equation (7.9), the CCFF obtain P_9 : $RN | \equiv U_{ser_i} | \equiv LCN_i \overset{s_k}{\leftrightarrow} RN$. This validates our target in **Ojective₃**. Finally, from H_{M6}, P_9 and Equation (7.8), the CCFF eventually achieves P_{10} : $RN | \equiv LCN_i \overset{s_k}{\leftrightarrow} RN$. This validates our target in **Ojective₄**.

Provided the aforementioned proved objectives $\langle$**Objective₁** − **Objective₄**$\rangle$, the proposed framework states that it uses a common s_k to cache data; and hence the proposed framework achieves the proper data fidelity and trust.

VI. CONCLUDING REMARKS

This work investigates the most appropriate Edge caching approach which can cope with the advances we are experiencing nowadays in the cloud/IoT era. There have been several attempts so far in the literature to come up with an efficient and secured caching approach which can enhance the requested/processed data accessibility and availability. However, these attempts suffer from critical aspects in Edge computing, including data latency, availability, security and scalability. Consequently, an IoT-based Edge framework, is recommended to be used, where the sensed data is mostly processed/requested at the edge of the network. In this framework a new component, called Local Cognitive Node (LCN)/Edge device, is introduced to retain information about data popularity and other security parameters that affect data availability and network scalability. It addresses the need for the delay-tolerant caching in edge networks. And it maximizes the gain of the data publishers by reducing their load.

REFERENCES

[1] B. Ahlgren, C. Dannewitz, C. Imbrenda, D. Kutscher, and Börje Ohlman. "A survey of information-centric networking," IEEE Communications Magazine, vol. 50, no. 7, pp. 26–36, 2012.

[2] P. Bellavista, A. Corradi, Alessandro Zanni, "Integrating Mobile Internet of Things and Cloud Computing towards Scalability: Lessons Learned from Existing Fog Computing Architectures and Solutions", 3rd Int. IBM Cloud Academy Conf. (ICA CON), 2015.

[3] IBM | A Smarter Planet | Smarter Cities. [Online]. Available: http://www.ibm.com/smarterplanet/us/en/smarter_cities.

[4] F. Al-Turjman, "Cognition in Information-Centric Sensor Networks for IoT Applications: An Overview", *Ann. Telecommun.*, pp. 1–16, 2016. doi: 10.1007/s12243-016-0533-8.

[5] F. Bonomi. Connected vehicles, the internet of things, and fog computing. VANET 2011, 2011.

[6] K. Shenai and S. Mukhopadhyay, "Cognitive sensor networks," in Proc. IEEE 26th Int. Conf. Microelectronics (MIEL), May 2008, pp. 315–320.

[7] G.T. Singh and F.M. Al-Turjman, "A Data Delivery Framework for Cognitive Information-Centric Sensor Networks in Smart Outdoor Monitoring", *Elsevier Computer Communications*, vol. 74, no. 1, pp. 38–51, 2016.

[8] P. Bellavista, A. Corradi, E. Magistretti. "REDMAN: An optimistic replication middleware for read-only resources in dense MANETs", Elsevier Pervasive and Mobile Computing, Vol. 1, No. 3, pp. 279-310, 2005.

[9] Docker. Available online at: https://www.docker.io

[10] Kura. Available online at: https://eclipse.org/kura.

[11] P. Bellavista and A. Zanni, "Feasibility of Fog Computing Deployment based on Docker Containerization over Raspberry Pi", *In Proc. of the ACM Int. Conf. on Distributed Computing and Networking*, Hyderabad, India, 2017.

[12] M. Burrows, M. Abadi, R. Needham, A logic of authentication, ACM Trans. Comput. Syst. Vol.8, no. 1, pp.18–36, 1990.

[13] Mozy-Backup Times: http://support.mozy.com/articles/en_US/Documentation/mozy-c-lotsofdata-howlong-faq

[14] Evernote pro. https://blog.evernote.com/blog/2016/09/13/evernotes-future-cloud/

[15] F. Al-Turjman, A. Alfagih, W. Alsalih, and H. Hassanein, "A delay-tolerant framework for integrated RSNs in IoT", Elsevier Computer Communications Journal, vol. 36, no. 9, pp. 998–1010, May, 2013.

[16] F. Al-Turjman, H. Hassanein, and M. Ibnkahla, "Efficient deployment of wireless sensor networks targeting environment monitoring applications", Elsevier: Computer Communications Journal, vol. 36, no. 2, pp. 135–148, Jan. 2013.

[17] A. Chankhunthod, P. Danzig, C. Neerdaels, M. Schwartz, and K. Worrell. A Hierarchical Internet Object Cache. In Proc. of USENIX, 1996.

[18] M. Gritter and D. R. Cheriton. TRIAD: A New Next-Generation Internet Architecture. Stanford University, July 2000.

[19] W. K. Chai, D. He, I. Psaras and G. Pavlou "Cache "Less for More" in Information-Centric Networks (Extended Version)", *Elsevier Computer Communications*, vol. 36, no. 7, pp. 758–770, 2013.

[20] S. Eum, et. al., "CATT: Cache Aware Target Identification for ICN", *IEEE Communications Magazine,* vol. 50, no. 12, 2012.

[21] A. Hail, M. Amadeo, A. Molinaro, S. Fischer, "Caching in Named Data Networking for the Wireless Internet of Things", *In proc. of the Int. Conf. on Recent Advances in Internet of Things (RioT)*, 2015.

[22] P. Radoslavov, R. Govindan, and D. Estrin, "Topology-informed internet replica placement," *Proceedings of WCW'01: Web Caching and Content Distribution Workshop,* 2001.

[23] S. Bhattacharjee, K.L. Calvert, E.W. Zegura, "Self-organizing widearea network caches", *In IEEE Infocom,* pp. 752–757, 1998.

[24] X. Vasilakos, V. Siris, G. Polyzos, and M. Pomonis, "Proactive Selective Neighbor Caching for Enhancing Mobility Support in Information-Centric Networks", *In Proc. of the ICN workshop on Information-centric networking,* New York, USA, pp. 61–66, 2012.

[25] M. Z. Hasan, et. al., "Optimized Multi-Constrained Quality-of-Service Multipath Routing Approach for Multimedia Sensor Networks", *IEEE Sensors Journal,* vol. 17, no. 7, pp. 2298–2309, 2017.

[26] I. Psaras, W. K. Chai, G. Pavlou, "Probabilistic In Network Caching for Information Centric Networks" *In Proc. of the 2nd edition of the ICN workshop on Information-centric networking,* pp. 55–60, 2012.

[27] V. Sourlas, P. Flegkas, L. Gkatzikis and L. Tassiulas "Autonomic Cache Management in Information Centric Networks", *In Proc. of the IEEE Network Operations and Management Symposium (NOMS),* 2012.

[28] F. Al-Turjman, "Cognitive Routing Protocol for Disaster-inspired Internet of Things", *Elsevier Future Generation Computer Systems,* 2017. doi: 10.1016/j.future.2017.03.014

[29] Hail M. A., Amadeo M., Molinaro A., Fischer S.: On the Performance of Caching and Forwarding in Information-Centric Networking for the IoT. 2015.

[30] J. Li, H. Wu, B. Liu, X. Wang, Y. Zhang, and L. Dong, "Popularitydriven coordinated caching in named data networking," pp. 200–211, 2012.

[31] K. Cho, M. Lee, K. Park, et al., "WAVE: Popularity-based and Collaborative In-network Caching for Content-Oriented Networks", *In Proceedings of INFOCOM WKSHPS,* pp. 316–321, 2012.

[32] J. Carbo, J.M. Molina and J. Davila, "Trust Management Through Fuzzy Reputation," *Int. Journal of Cooperative Information Systems,* vol.12, no.1, pp. 135–155, 2003.

[33] Z. Ming; M. Xu; D. Wang "Age-based Cooperative Caching in Information-Centric Networks", *Int. Conf. on Computer Communication and Networks (ICCCN),* 2014.

[34] W. Yaogong, K. Lee, B. Venkataraman, et al. "Advertising Cached Contents in the Control Plane: Necessity and Feasibility", *In Proc. INFOCOM Workshop on computer communications,* 2014.

[35] S. Borst, V. Gupta, A. Walid, "Distributed Caching Algorithms for Content Distribution Networks", *In Proc. of the IEEE INFOCOM,* 2010.

[36] C. Fricker, P. Robert, J. Roberts, N. Sbihi, "Impact Of Traffic Mix On Caching Performance In a Content-Centric Network", In *INFOCOM Workshops,* pp. 310–315, 2012.

[37] F. Al-Turjman, H. Hassanein, and M. Ibnkahla, "Towards prolonged lifetime for deployed WSNs in outdoor environment monitoring", Elsevier Ad Hoc Networks Journal, vol. 24, no. A, pp. 172–185, Jan., 2015.

[38] A. Al-Fagih, F. Al-Turjman, W. Alsalih and H. Hassanein, "A priced public sensing framework for heterogeneous IoT architectures," IEEE Transactions on Emerging Topics in Computing, vol. 1, no. 1, pp. 135–147, Oct. 2013.

[39] F. Al-Turjman, H. Hassanein, S. Oteafy, and W. Alsalih, "Towards augmenting federated wireless sensor networks in forestry applications", Springer: Personal and Ubiquitous Computing Journal, vol. 17, no. 5, pp. 1025–1034, June, 2013.

[40] F. Al-Turjman, H. Hassanein, and M. Ibnkahla, "Quantifying connectivity in wireless sensor networks with grid-based deployments", Elsevier: Journal of Network & Computer Applications, vol. 36, no. 1, pp. 368–377, Jan, 2013.

8

SECURE ACCESS IN eHEALTH SPACES[1]

Contents

[1] Fadi Al-Turjman is with Antalya Bilim University, in Turkey (e-mail: fadi.alturjman@antalya.edu.tr).

Sinem Alturjman is with Antalya Bilim University, in Turkey (e-mail: s.oncel_92@hotmail.com).

Previously published in F. Al-Turjman, and S. Alturjman, "Context-sensitive Access in Industrial Internet of Things (IIoT) Healthcare Applications", *IEEE Transactions on Industrial Informatics*, vol. 14, no. 6, pp. 2736-2744, 2018.

Abstract

Industrial Internet of Things (IIoTs) is the fast growing network of interconnected things that collects and exchange data using embedded sensors planted everywhere. Several IIoT applications such as the ones related to healthcare systems are expected to widely utilize the evolving 5G technology. This 5G-inspired IIoT paradigm in healthcare applications enables the users to interact with various types of sensors via secure wireless medical sensor networks (WMSNs). Users of 5G networks should interact with each other in a seamless secure manner. And thus, security richness is highly coveted for the real time wireless sensor network systems. Asking users to verify themselves before every interaction is a tedious, time-consuming process that disrupts inhabitants' activities and degrades the overall healthcare system performance. To avoid such problems, we propose a Context-sensitive Seamless Identity Provisioning (CSIP) framework for the IIoT. CSIP proposes a secure mutual authentication approach using *Hash* and *Global Assertion Value* to prove that the proposed mechanism can achieve the major security goals of the WMSN in a short time period.

Keywords

Mutual authentication, security, wireless medical sensor network

I. INTRODUCTION

Following the remarkable success of 2G and 3G mobile networks and the fast growth of 4G, the next generation mobile networks, including the 5G and Industrial Internet of Things (5G/IIoT) was proposed aiming to provide endless networking capabilities to mobile users. Industrial Internet of Things (IIoT), also known as industrial internet, brings together smart machines, innovative analytics, and people at work. It is an interconnection of many devices through a diverse communication system to bring forth a top-notch system capable of monitoring, collecting, exchanging, analyzing, and delivering valuable information. These systems can then help manage smarter and faster business resolutions for industrial companies. As the author in [1] puts it, IIoT is more advanced than commercial IoT, simply due to the dominance of the connected sensors in the industrial platform. Sensor interface is a key factor in industrial data collection, the author in [2] states that the present connect number, sampling rate, and signal types of sensors, are highly restricted by the sensing device.

To advance beyond the traditional mobile networks, intelligence, and secure network access need to be diffused, empowering even the smallest connected sensors. IIoT can revolutionize the ubiquitous computing with multitude of applications built around various types of "smart" sensors enabled with intelligence and machine learning techniques. Smart sensors evolve our knowledge and conception of the world as a hyper-connected environment that has raised new requirements for making the IIoT ready for large scale deployments. This new paradigm is all about "sensing", an evolution, or maybe revolution, that will take the mobile users into new territories. Secure communication and sensing techniques enable a participatory approach for achieving integrated solutions and creating novel applications related to industry and especially to healthcare. With the proliferation of intelligent identification methods, sensors are expected to lead further innovation in IIoT. Sensors play an undeniably key role in driving the IIoT revolution by incorporating 5G networks that provide functions such as data conversion, spectrum sensing, digital processing and communication to external devices. 5G is expected to be more than a new generation of mobile communications. Instead, it is already considered as the unifying fabric that will connect billions of devices in some of the fastest, most reliable and most efficient ways possible. Of course, the impact of such an enabling technology is expected to be revolutionary. The new infrastructure for communication is expected to transform the world of connected sensors and reshape industries. Such a revolution would of course require research and development for the co-existence and device inter-operability for sensors with 5G networks. Integrated IIoT is an intelligent communication network capable of acquiring knowledge about its users and its surroundings, and uses such knowledge to help users achieve their goals in a context-based manner. This definitely improves users' quality of life, and helps in optimizing and controlling the dramatically increasing consumption rates of resources in IIoT environments. Inhabitants (users) of IIoT environments could be people, devices, services, or systems, occupied with smart enabling technologies such as RFIDs, sensors, 5G smartphones, etc. in varying applications in our daily life. A key application for the 5G-based IIoT is the eHealthcare, which aims at maintaining the patient's medical information in electronic environments such as the Cloud via up to date telecommunication paradigms. In eHealthcare application system, Wireless Medical Sensor Networks (WMSNs) have become a prominent technology for Wireless Sensor Networks (WSNs) [2]. For the early diagnosis, the hospitals have developed several tiny medical sensors that are used to sense the patients' body to collect the health data, such as heart beat, blood pressure, temperature and so on. Besides, these sensing data are broadcasted via the profession handheld devices, namely PDA and smart phone to-do further analysis. Various medical research communities have begun their patient health monitoring using WMSNs [1][3][4][5]. Therefore, user authentication scheme is necessitated to protect the illegal access in the medical information system.

WSNs' have recently been a pioneer for the source of data unification. Further, it can be able to provide an executable solution to the critical mission applications. The deployment of WSNs' can be a proviso for the various mission critical environments such as surveillance in military, fire detection in forest, medical health care system and in monitoring a wild life. Recently, several research works have proven that the heterogeneous sensor networks have better network performance, more reliable, scalable, transparent, load-balance, lifetime and cost-effective [6]-[9]. Hence, the heterogeneous types of the sensor networks have been a salient system for the practical related real time applications. Though it has more prominent features [6]-[9], it still requires a promising user authentication scheme to offer mutual authenticity, secure session key sharing and privileged-insider resiliency. Thus, the real time mission-critical application necessitates a secure cum mutually reliable authentication scheme to protect the network system from the anomalies. In the past, numerous user authentication protocols have been proposed for the heterogeneous sensor network system [10]-[13]. Since the existing mechanisms focus on unilateral user authentication, it is thus not able to deduce whether the connection is legitimate or not; further the trustworthy between the communication entities becomes zero. Le et al. in [14] presented a mutual-authentic based secure authentication using elliptic-curve cryptography (ECC). However, the Le et al. scheme is susceptible to information leakage attack. In 2009, Das in [15] discovered a two-factor user authentication scheme for WSNs. Since the Das scheme uses a one-way hashing to encrypt and decrypt the sensed data, several authentication schemes [16]-[19] prove that the Das scheme is insecure to offline password guessing, key-impersonation, privileged-insider and gateway (node) bypassing attack. Lately, He et al. [20], Yoon et al. in [21] and Chuang et al. in [22] have proposed the user authentication schemes, where the pitfalls are: 1. The computational cost is expensive; and 2. The security weaknesses, like mutual authentication and session-key establishment and vulnerabilities, like privileged-insider attack are intangible.

To address the above issues in the real time based WSNs' application, this paper proposes a secure cum mutual authentication scheme that draws a prominent feature of the symmetric cryptosystem to provide strong mutual authenticity, secure-key agreement and resilient to privileged-insider attack to use in wireless multimedia medical sensor network (WMSN) system. We propose CSIP, a secure context-sensitive seamless multi-modal identity provisioning framework for smart environments. CSIP builds an encrypted and compressed 360-Degree Inhabitant Profile [23][24] by using his activities' history and usage patterns of the environment's resources, based on that, it can build Disposable Customized Virtual Inhabitant Profile (DCVIP) then, it creates an identity proxy to perform the verification required during the interaction. Moreover, the paradigm is lightweight, and hence perfect for 5G-based technologies. We can divide the CSIP into two parts, client-based part and cloud-based part. The first part considers the client and it is responsible for gathering inhabitants' access data into blocks, then compresses these blocks and sends it to the cloud-side. The second part is related to the cloud, it receives inhabitants' data from the first part, applies deep analysis to gather the information required for the interaction, and then classifies and saves this information. We believe CSIP reduces the risks for identity theft/loss, as it does not depend on static identifiers, and it takes into consideration the dynamics of the inhabitants' behaviors as well as their access and usage patterns to handle incoming requests. Extensive simulation results and security analysis show that CSIP is practical as it provides acceptable performance when compared with a basic static identity proxy approach. In the following, we overview the attempts related to this system in the literature.

In eHealthcare, WSNs represent the infrastructure, which extract the information of smart sensing object [25]. WSNs are one of the essential components of the infrastructures employed for smart eHealthcare in IoT-based applications. Recently, security issues in WSNs have gained much attention of the researchers not only to satisfy the security properties of authentication and key agreement (AKA) protocol but also to mitigate the computation and communication cost of the system. For the achievement of minimum overhead, several lightweight authentication schemes have been proposed [26],[27]-[32]. Watro et al. proposed the lightweight two-factor user authentication based on RSA cryptosystem for WSNs. However, the Watro et al. scheme [27] is vulnerable to replay, denial of service and key impersonation attacks. Wang et al., [28] presented a lightweight user authentication scheme for WSNs, which only demands the computation of a hashing function. Later on, Srinivas et al. [30] show that the Wang et al. scheme [28] is vulnerable to stolen verifier and many logged-in users with the same login identity attack. Tseng et al. [31] improved the version of Wong et al., which does not offer mutual authentication between the base-station and sensor-node. To overcome the security weakness of mutual authentication, Lee [32] presented a novel password based dynamic user authentication scheme, which also fails to satisfy mutual authentication between the base-station and the sensor-node.

Meanwhile, Cloud-based solutions provide significant long- and short-term benefits over an internal computing grid. Over the long-term, Cloud storage, analysis and archival are scalable for the streaming, real-time data volume that is expected to be generated by users over an extended period of time. In the short-term, Cloud solutions provide more cost-effective deployment. Maintenance costs are also optimized, since the Cloud infrastructure is kept up to date by the Cloud services providers. The proposed CSIP framework addresses the identity verification problem by only accumulating data related to the user's access to resources. Collecting such information does not necessitate the patients to do extra disorderly activities in the process of confirming their individuality. Moreover, only activity types are used in the confirmation process, instead of the activities themselves, which makes the proposed framework less conspicuous than biometric and context-aware methods [9][10]. As the data needed by the framework is composed anyhow and no further information is obligatory, the verification choices are made more rapidly.

The rest of the paper section is devised as follows. Section II describes the system models. Section III presents the CSIP secure mutual authentication scheme for the real time eHealthcare application. Simulation results and analysis of the security properties are discussed in Section IV. Finally, Section V concludes the paper work.

II. SYSTEM MODELS

This section will discuss the system model of WMSN and related assumptions.

A. Network Model

The considered network model for the proposed WMSN is illustrated in Figure 8.1 in which the patient may use medical sensor / smart phone to sense and update health metrics, such as current blood pressure, body movement, heart bit rate, pulse rate and so on for the doctor / medical technical expert.

The doctor / medical expert may access the current health information of the patients via smart phone system, such as PDA and laptop.

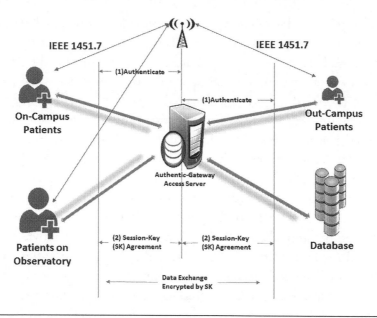

Figure 8.1   Network model of the proposed WMSN.

The WMSN system model has three actors, namely patient, medical expert, authentic-gateway access.

Since the patient and medical experts share the confidential information over an Ad hoc network, this paper thus interpolates an authentic-gateway system on the Ad hoc network to provide proper user authentication, secure-session key sharing and minimum computation overload as the information is sent fast with little or no slow human intervention. Additionally, it is user friendly and resistant to potential attack of a privileged-insider, from unauthorized personal access to confidential information, by providing a secure gateway between the physician and the patient.

In addition, a smart PnP sensor like IEEE 1451.7 (incorporated with the medical sensor) is used to define the hardware and software transducer interface (sensor or actuator). The specific objective of IEEE 1451.7 [17] is the sensor security that is interfaced between sensor and user, like patient / medical expert to achieve the features of mutual authentication and session-key establishment. And thus, it is preferred for security operations and secure data (multimedia) transmissions. To simplify the sensor descriptive characteristics, a smart sensor like medical sensor is incorporated as a Transducer Electronic Data Sheet (TEDS). The main purposes of this technology are:

1. Assisting the host-device to identify the sensor related parameters;
2. Providing the wireless connection to the host-device provided self-descriptive sensor parameters.

III. PROPOSED CSIP APPROACH

In this section, this paper presents a secure-cum-mutual user authentication scheme for the WMSN system using medical-sensor / smart phone and the prime objective of this scheme is to ensure that the medical transmission of information between the patient and doctor are mutually authenticated or not. To execute the authentication scheme, the following assumption is considered: 1) The authentic-gateway is believed to be a trustworthy node; 2) Two long-term master key should be used between the entities (based on existing identity-based key agreement protocols such as SCK-1, and SCK-2 [25]); and 3) A long-term secret-session key should be shared to ensure mutual authenticity $ssk_{a-gw} = H(S_N \oplus ID_{gw})$.

A. Three-Phase CSIP

The proposed scheme is comprised of three phases: system registration, system login and system authentication.

Phase I - System Registration   In this phase, the medical expert enters the credentials into the authentic-gateway system and its related execution flows are as follows:

Step 1: The medical expert chooses M_{id} and s_k and then he / she submits into the authentic-gateway node via secure channel.

Step 2: Upon receiving the medical expert M_{id} and s_k, the authentic-gateway determines the followings: $C = E_J[M_{id} \| ID_{gw}]$ and $N_i = H(M_{id} \oplus s_k \oplus S_{key})$.

Step 3: Thenceforth, the authentic-gateway provides a secure-ware to the medical expert with the configuration of the following parameter $\{H(.), C, N_i, S_{key}\}$. Herein, S_{key} is a long-term gateway secret key that is securely bound between the entities.

Phase II - System Login and User-Authentication Phases This phase may be invoked, when the medical expert visits the patients' ward and wishes to review the current information status of the patient. To access such confidential data, the medical expert should enter a proper long-term secret key into the smart phone system. Upon receiving the login-request, the authentic-gateway verifies the long-term secret key into the system database to execute the following operations: $N_i^* = H\left(M_{id} \oplus s_k \oplus S_{key}\right)$; Compare: $N_i^* = N_i$; compute: $H\left(M_{id}\right)$ and $CID_i = E_k[H\left(M_{id} \parallel M \parallel S_N\right]$; and eventually, generate $\{CID_i, C, T'\}$, then sends to the authentic-gateway node. Herein, M is a random nonce determined by the medical expert to establish secure session-key. Upon receiving the medical expert's message, the authentic-gateway executes the following tasks to authenticate his/her access-request:

Step 1: Initially, the authentic-gateway node validates the current access time T_c: validate whether $\left(T_c^{''} - T_c^{'}\right) \geq \Delta T_c$, if the expression holds, then the authentic-gateway node refuses the access-request and terminate the process. Otherwise, the authentic-gateway executes the further steps. Herein, $T_c^{''}$ is the current request-time of the authentic-gateway and ΔT_c is the delay time interval.

Step 2: After the successful validation, the authentic-gateway executes the following tasks: Compute: $D = D_J\left[M_{id}^* \parallel ID_{gw}^*\right]$ from the M_{id}^* and ID_{gw}^*; Compute: $H\left(M_{id}^*\right)$; Compute: $D_k[CID_i] = E_k[H\left(M_{id} \parallel M \parallel S_N\right]$ from the M_{id}^*, M and S_N. Compare $H\left(M_{id}^*\right) = H\left(M_{id}^{'}\right)$ and $ID_{gw}^* = ID_{gw}$; if the condition is satisfied, then the request is authentic; Otherwise, terminate the rest of the process; Compute $V_i = E_{SK_{gw}}\left[M_{id} \parallel S_N \parallel M \parallel T_c^{'''}\right]$; generate the request-message $\{V_i, T_c^{'''}\}$ and then sends the request-message to the nearby medical sensor / access point wherein the medical expert is available to access the patient info.

Step 3: Upon receiving the authentic-gateway message, the medical sensor / smart phone executes the following tasks: S_N verifies the time $T_c^{'''}$: validate whether $T_c^{''''} - T_c^{'''} \geq \Delta T_c$, if the validation is successful, the medical sensor node rejects the authentic request-message of the authentic-gateway and terminate the further process. Herein, $T_c^{''''}$ is the current execution time of the medical sensor node and ΔT_c is the delay time interval.

Step 4: After the successful validation, the medical sensor node executes the following tasks: Compute $D_{Sk-gw} = D_{gw}\left[M_{id}^* \parallel S_N^* \parallel ID_{gw}^* \parallel M^*\right]$ from the $M_{id}^*, S_N^*, ID_{gw}^*, M^*$;

Compare $S_N^* = S_N$, if the condition holds, then terminate the request-message of authentic-gateway node; otherwise continue the further process; S_N computes the secure-session key $ss_k = H\left(M_{id}^* \| S_N \| M^*\right)$; Compute: $L = E_{ssk}\left[(S_N \| M^* \| T_c^*\right]$; Generate the response-message $\{L, T_c^*\}$ and then sends the response-message to the medical expert.

Step 5: Upon receiving the response-message from the medical sensor node, the medical expert executes the following tasks: Medical expert validates the time interval T_c^*. Validate, whether $T_c^{**} - T_c^* \geq \Delta T_c$, if the condition is satisfied, then refuses the request and terminate the process. Otherwise, continue the rest of the process. Herein, T_c^{**} is the current process time of the medical expert's system and ΔT_c is the delay time interval.

Step 6: After the successful validation, the medical experts' system executes the following tasks: Compute: $SK = H\left(M_{id} \| S_N \| M\right)$; Decrypt: Using L, to obtain the valid SK, S_N^* and M^*; Compare: $S_N^* = S_N$ and $M^* = M$, if the condition is satisfied, then the secure session key has been established between the medical sensor and experts successfully; Otherwise not established successfully.

Phase III - Session-Key Update Phases The session-key update is called forth while U_{ser_i} wishes to update his / her password. The working procedure for secret-update is as follows:

Step 1: U_{ser_i} puts his / her smart-card in the terminal to enter his / her credentials, namely M_{id} and s_k.

Step 2: Upon the credentials validation and verification, the smart-card carries out the operation of $N_i' = H\left(M_{id} \oplus s_k \oplus S_{key}\right)$ to compare with $N_i' = N_i$. If the comparison is successful, then the rest of the operation will be proceeded. Otherwise, the operation will be aborted.

Step 3: If validation is successful, U_{ser_i} is asked to enter a new secret-key s_k^{new}.

Step 4: Compute $N_i^{new} = H\left(M_{id} \oplus s_k^{new} \oplus S_{key}\right)$ to replace N_i with N_i^{new} in the smart-card.

B. Cloud-Based CSIP Architecture

The architecture of the cloud-based side of CSIP is depicted in Figure 8.2. Figures show the client-based CSIP consists of about thirteen main modules described as follows:

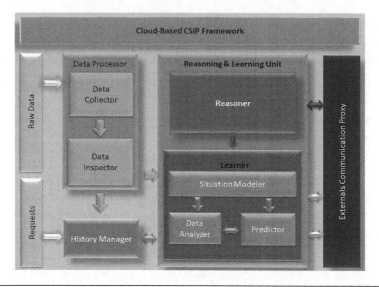

Figure 8.2   Cloud-based CSIP Architecture.

CSIP Clients Communication Proxy: the smart device might not be supporting the HTTP protocol (such as Contiki devices, and devices adopting CoAP protocol), hence this module must do the mapping between the protocols, and synchronize transferring the blocks with the cloud-based side.

Data Processor: It is the module responsible for collecting the data blocks and requests from the CSIP clients, then pass them to the history manager and reasoner modules. The data processor consists of a data collector and a data inspector. The data collector collects the data from different CSIP clients, while the inspector checks if the proper encryption and compression techniques are adopted, also it performs pre-processing steps on the collected data before passing it to the reasoner and history manager modules.

History Manager: It is the module responsible for storing the inhabitants' data blocks and requests, so they can be accessed by the reasoning module for learning purposes.

Reasoning & Learning Unit: It is the module responsible for detecting and predicting the inhabitants' usage and access patterns. This module should apply different on-line learning techniques to be able to model inhabitants' situations based on the coming inputs and previous history of inhabitants' interactions. The reasoner consists of a data analyzer, a predictor, and a situation modeler. The data analyzer applies different data mining techniques to extract the access and usage patterns. The predictor uses the given data to generate the expected inhabitant behavior in the detected context. The data analyzer and predictor send their results to the situation modeler to create a model for the current inhabitant situation. If the predictor model and the data analyzer results are not matching, the situation is considered a conflict and the reasoner has to make a decision to resolve such conflict.

IV. PERFORMANCE EVALUATION

In this section, the proposed CSIP framework is evaluated and compared with related authentication schemes. To evaluate the various cryptographic operation, an extensive verification is done using MIRACLE C/C++ library with the system features of 32-bit Windows 7 Operating Systems and Microsoft Visual C++ 2008 software edition. To examine realistically, the execution time of symmetric key encryption/decryption $(AES-128)$, elliptic-curve point scalar multiplication over finite-field f_p and $SHA-1$ hashing function are set as $T_{ED} \approx 0.1303\ ms$, $T_M \approx 0.7.3529\ ms$ and $T_{SH} \approx T_{MH} \approx 0.0004\ ms$ as referred in [26]. To better understand the evaluation criteria of communication cost, some notation is defined as follows:

- T_{SH} is defined as the execution time of one-way secure hashing function $H_2(.)$. T_{MH} is defined as the execution time of one-way point to map hashing function $H_1(.)$.
- T_P is defined as the computation time of bilinear pairing function.
- T_A is defined as the execution time of one-point additional operational function.
- T_{ED} is defined as the execution time of encryption and decryption algorithmic function.
- T_M is defined as the execution time of elliptic-curve scalar multiplication function.

This subsection will explain the requirements in detail that are as follows:

Mutual Authentication: In the mutual authentication, the two communication parties authenticate each other for the purpose of secure communication establishment. In the proposed scheme, the patient and doctor authenticate each other using $ssk_{a-gw} = H\left(S_N \oplus ID_{gw}\right)$. In the system login and authentication phase, the authentic-gateway access server authenticates the patient / doctor using the computation of $D = D_J\left[M_{id}^* \parallel ID_{gw}^*\right]$ and $D_k\left[CID_i\right] = E_k\left[H(M_{id} \parallel M \parallel S_N\right]$ to verify whether the patient / doctor satisfies the conditional expression of $V_i = E_{SK_{gw}}\left[M_{id} \parallel S_N \parallel M \parallel T_c''\right]$ to access the data being transmitted. Though the adversary intercept the login request-message of either patient or doctor and wishes to forge as a legitimate authentic-gateway server, the adversary cannot compute the parameters' like $\left\{H(.), C, N_i, S_{key}\right\}$. Subsequently, the adversary cannot send a valid response-message to the authentic-gateway server. Hence, the proposed scheme asserts that it holds the security property of mutual authentication.

Secret Key Generation: In the proposed scheme, the patient and doctor can share a secure session-key via an authentic-gateway access after the successful execution of authentication phase. By then, the patient and doctor can exchange the real-time data securely using the establishment of secure-session key which is used to encrypt the real-time data gathered by the smart medical sensor. The secure session-key is determined from the computation of $ssk_{a-gw} = H\left(S_N \oplus ID_{gw}\right)$ and validated from the computation of $V_i = E_{SK_{gw}}\left[M_{id} \parallel S_N \parallel M \parallel T_c''\right]$. Since the parameter of T_c changes over a period of time, thus the different set of a session – key will be used to authenticate the session. Hence, the proposed scheme asserts that it holds the security property of the session - key agreement.

Resilient to Privileged-Insider Attack: The proposed authentication scheme never sends the parameters, such as $\left\{H(.), C, N_i, S_{key}\right\}$ to the authentic-gateway access server as the plaintext. The user, like patient / doctor sends the parameter as $H\left(S_N \oplus ID_{gw}\right)$, thus the authentic-gateway access cannot acquire the users' secret-key without the knowledge of $H\left(M_{id}\right)$ and $CID_i = E_k\left[H(M_{id} \parallel M \parallel S_N\right]$; since it is highly secured. Moreover, the expression $H\left(S_N \oplus ID_{gw}\right)$ is eventually verified using $V_i = E_{SK_{gw}}\left[M_{id} \parallel S_N \parallel M \parallel T_c'''\right]$ to authenticate the access of the session to the patient / medical expert. Thus, the adversary cannot compute a valid session-key without the proper deduction of S_N, ID_{gw}. Hence, the proposed scheme asserts that it is resilient to the attack of a privileged - insider.

Resilient to Replay Attack: Assume that the adversary uses a previous message to authentic-gateway $\{CID_i, C, T'\}$, medical sensor node $\{V_i, T_c'''\}$ and user $\{L, T_c^*\}$ to gain the access privilege. As the messages are validated using fresh timestamp, the adversary cannot be succeeded with the entries of previous message. Thus, the proposed scheme assert that it is resilient to the relay attack.

Resilient to User Masquerading Attack: As an instance, the adversary has forged a message of login $\{CID_i, C, T'\}$. So, he / she tries to login with an updated message $\{CID_i^{new}, C, T'\}$ to duplicate the context of the user identity CID_i^{new}. However, the context identity CID_i^{new} cannot decrypt the original message using $\left[M_{id}^* \parallel S_N^* \parallel ID_{gw}^* \parallel M^* \right]$. Thus, the proposed scheme claims that it can resist the user masquerading attack.

Resilient to Secret Gateway Guessing Attack: The proposed scheme has three different master keys, namely k, M and J. These keys are not transmitted as the plaintext; and thus the proposed scheme asserts that it is resilient to the secret gateway guessing attack.

By the above procedure, the medical expert and sensor node can authenticate one another to access the wireless multimedia medical sensor network (WMSN). Eventually, the medical expert / doctor can access the private info of the patient from the medical sensor node / smart device via authentic-gateway node. Since the secure session-key is tightly bound between the entities, the proposed CSIP mechanism can achieve the security goals, namely mutual authentication, session-key establishment and resilient to privileged-insider, replay, user masquerading, and secret gateway guessing attack for the WMSN as shown in Table 8.1. Unlike other competitive baselines in the literature, CSIP achieves all the above mentioned security

Table 8.1 Comparison of Security Properties

	[20]	[21]	[22]	PROPOSED CSIP SCHEME
Mutual Authentication	Not Provided	Yes	Yes	Yes
Session-Key Agreement	Not Provided	Partial	Yes	Yes
Resilient to Privileged-Insider Attack	Not Provided	Yes	Partial	Yes
Resilient to Replay Attack	Not Provided	Yes	Yes	Yes
Resilient to User Masquerading Attack	Not Provided	No	No	Yes
Resilient to Secret Gateway Guessing Attack	Not Provided	No	No	Yes

goals. Table 8.2 shows the computational efficiencies of proposed authentication scheme. Since the proposed scheme has less computation cost, it mitigates the execution time of the phase to improve the performance of the WMSN in comparison with the other authentication schemes [20][21][22]. Moreover, we compare our proposed CSIP approach against the other authentication schemes in terms of number of required messages to deliver a specific amount of data (e.g., 512 bits), bandwidth utilization and overhead percentage as shown in Table 8.3. In this table, we assume three security levels per approach, which are L_1–L_3, respectively, and we assume that the corresponding length of the message authentication codes are 2/4/6 (in bits). Since our CSIP approach allows data with different levels of security requirement to be packed together while other approaches do not allow this, both the obtained number of messages and the bandwidth overhead are smaller in general for CSIP as compared to

Table 8.2 Computational Efficiency

	SYSTEM REGISTRATION		SYSTEM LOGIN AND AUTHENTICATION		
SCHEME	MEDICAL EXPERTS	AUTHENTIC-GATEWAY	MEDICAL EXPERTS	AUTHENTIC-GATEWAY	MEDICAL SENSOR NODE
Proposed CSIP Scheme	–	1H+1S	4H+2S	1H+3S	1H+2S
[20]	2H	6H+1S	7H+3T_E	3H+1S+1T_E	4H+1T_E
[21]	1H	(n+m) H	2H+1ECC	5H+2ECC	3H+1ECC
[22]	1H	(n) 2H	4H	8H+1E	8H+1E

Table 8.3 Comparison of Design Factors Under Varying Security Levels

	[20]			[21]			[22]			PROPOSED CSIP SCHEME		
	L_1	L_2	L_3	L_1	L_2	L_3	L_1	L_2	L_3	L_1	L_2	L_3
Number of messages	20	21	23	18	19	21	18	18	20	16	17	17
Bandwidth utilization (%)	20.19	21.5	22.3	19.67	20.12	22.6	21.93	22.6	24.1	23.67	24.12	25.6
Bandwidth overhead (%)	5.17	6.2	7.9	3.99	4.1	5.2	4.73	3.8	5.1	2.73	2.91	3.1

others. The reason is that compared with the bandwidth overhead brought by increasing the length of the message authentication code to guarantee the message security, the experienced bandwidth overhead by dividing data into different messages is much bigger. We can see from Table 8.3 also that along with the increasing of the security requirement, the obtained number of messages and the bandwidth overhead also increase. But as CSIP is based on learning approach, even when the level of security requirement reaches L_3, the bandwidth overhead is limited to 3.1%. And thus, CSIP solves the tradeoff problem, where the bandwidth utilization is improved and the requirements in both security and real-time are met.

We remark also that CSIP divides the protected resources in the smart environment into a sum of assertion/security levels (n). All the following tests which have been used in assessing the CSIP method presumed the total of assertion levels n equal to 3. The verge values of the assertion levels are set as follows; $g_{t1} = 0$, $g_{t2} = 0.33$, and $g_{t3} = 0.66$. Where $g_{ti} \in [0,1]$ and represents the assertion level threshold value of the ith assertion level that can be calculated based on the following equation:

$$g_{ti} = (i-1) * \frac{1}{n} \tag{8.1}$$

It imitates how assured the system must be about a user in order to state his/her identity before he/she is allowed to access the resource.

A. The Effect of Δg

Altering the parameter Δg marks how the system assurance is dispensed as shown in Figure 8.3. In this figure when Δg is set to values between 0.1–0.5, the scheme assurance increases gradually and needs at least 6 events for the global assertion value to reach the following threshold value. This situation would be useful in extremely protected locations. However, when Δg has to experience values such as 0.7 – 1.0, the scheme assurance increases much faster with less number of events to grasp the second level; this situation would be useful for more relaxed surroundings.

In Figure 8.4 the consequences of trying the identification for major events are examined. This figure shows big differences for the three

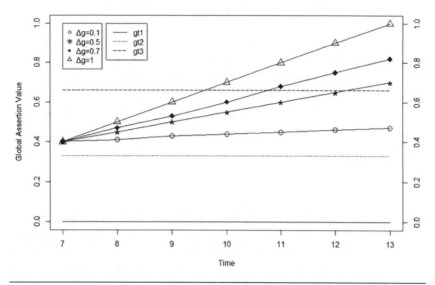

Figure 8.3　CSIP response for repeated events with altered Δg values.

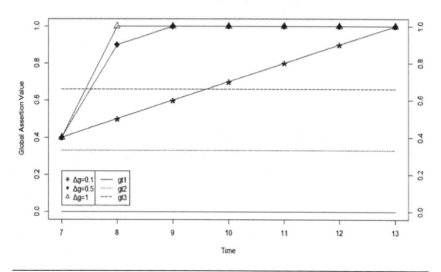

Figure 8.4　CSIP response for identifying major events with altered Δg settings.

examined Δg values (0.1, 0.5 and 1). In case of Δg value equal to 0.1, the assertion value increases slowly and necessitates almost three significant events to reach the next assertion level; this is considered to be undesirable as it indicates the CSIP scheme could not believe the identified event. For Δg equal to 0.5 and 1, the affirmation value rises faster and necessitates one major event to reach the equal effect, which is more preferable.

B. The Effect of Assertion Value on the CSIP Response

Picking diverse settings to denote the local assertion values for repeated events will yield alike responses for strict, moderate and relaxed surroundings per Figure 8.5. As depicted in the figure, selecting K_r values between 0.7 and 1.0 renders this type of event to have an identical effect to that of the major event type. Consequently, K_r is suggested to be assigned a value between 0.1 – 0.5.

To verify the security and suitability of the proposed secret key generation (SKG) scheme, we compare the SKG Rate (SKGR) of the proposed scheme based on private pilot with the approaches based on random beam-forming described in [33], which are based on a public pilot, using Monte Carlo simulations. We assume Rayleigh fading channels and the entries obey Gaussian distribution with zero mean and unit variance. In addition, we consider the receiver noises as white Gaussian noises with zero mean and unit variance, and a middle person, Eric, can estimate channels H_{AE} between Alice and Eric, and H_{BE} between Bob and Eric, while applying the SKG with a public pilot.

We first investigate how the location of the man-in-the-middle (MITM) attacker; Eric, can affect the SKGR when the transmitting powers of Alice, Bob and Eric are described by SNR = 10 and

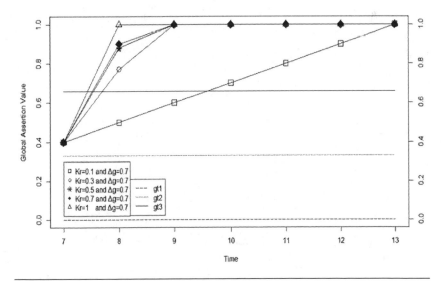

Figure 8.5 CSIP response for diverse repeated event values.

$N_A = N_E = 4$, and $N_B = 1, 2$, and 4. Where Alice, Bob, and Eric have N_A, N_B, and N_E antennas, respectively. We normalize the distance from Alice and Bob to Eric as 1, and assume that the channel gain between antenna i of the transmitter and antenna j of the receiver is $h_{ij} = \chi d_{ij}^{-l/2}$, where d_{ij} is the distance between i and j, $\chi \sim \mathcal{N}(0, \sigma^2)$, and $l = 2$ is the path-loss exponent. We assume that Eric moves along the wireless channels from Alice to Bob, the distance changes from 0.1 to 0.9 in the interval 0.05.

Figure 8.6 shows that the SKGRs rise at first, reach the maximum values, and then decrease with the increase of the distance between Alice and Eric, while there are some differences about location of the maximum; such that it reaches the maximum at 0.6, 0.5 and 0.45, respectively, when $N_B = 1, 2$, and 4. It indicates Eric can intercept more information when it is closer to Bob than Alice, and Eric moves to the middle location with the antenna count increase at Bob. Since both multi-antenna and high SNR can increase the SKGR, this result is expected.

Secondly, we investigate the security of the reconstitution wireless channels (RWC) in static or quasi-static environments (SQSE)

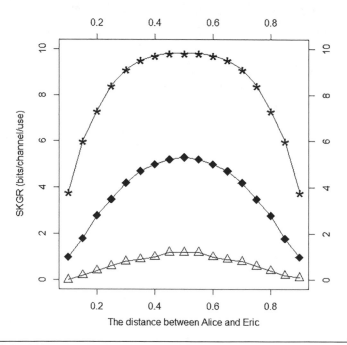

Figure 8.6 SKGR varies with the location of attacker.

by comparing the SKGR based on private pilot and reconstitution channels with that based on the random beam-forming described in [33] when $N_A = N_E = 4$, and $N_B = 1$, and $N_A = N_E = N_B = 4$, respectively, which are corresponding to the MIMO and MISO scenarios, respectively. We explore the SKGR for three cases under the MITM attacks: 1) no attacker; 2) based on random beam-forming (RBF) described in [33]; and 3) based on the proposed private pilot. The former approaches use public pilot to estimate the wireless channels.

Figure 8.7 shows that the SKGRs based on the private pilot and reconstitution channels rise with the increasing of the SNR of the transmitter, which is close to the SKGR without attacker. However, the SKGR based on random beam-forming with public pilot described in [33] are almost equal to zero under the MITM attacks, which shows that the MITM attack is a serious threat to SKG.

V. CONCLUSION

In this paper, a secure-CSIP mutual authentication framework has been proposed for the health care system using wireless multimedia medical sensor network. In addition, this paper has exploited the two-factor (namely medical expert and sensor node) strategy for the mitigation of computational cost and fulfillment of WMSN security goals, such as session-key agreement and resilient to privileged-insider, replay, user masquerading and secret gateway

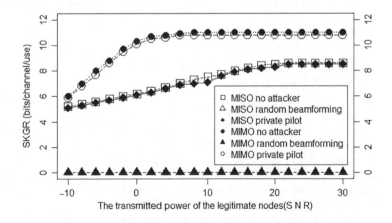

Figure 8.7 SKGR varies with SNR of transmitter.

guessing attack. The proposed authentication meets the demand of platform adaptability; and thus suitable for all the real time mission critical applications systems. Besides, the proposed scheme shows that it has less computational overhead to improve the performance efficiency of the systems. Moreover, it solves the natural tradeoff between security level and the added communication overhead, where the bandwidth utilization is improved and the requirements in both security and real-time are met. It considers several security threats including privileged-insider, replay, user masquerading, secret gateway guessing, and man-in-the-middle (MITM) attacks. Furthermore, we showed a real-life use case for CSIP adoption while investigating critical design factors. Finally, the investigated use case scenarios show that CSIP is successful in verifying user identity with high degree of reliability. It is particularly effective in certain environments that has a combination of different security level requirements and the user is authenticated automatically based on history/learning mechanisms.

REFERENCES

[1] B. Karschnia. "Industrial Internet of Things (IIoT) benefits, examples | Control Engineering", Controleng.com, 2017. [Online]. Available: http://www.controleng.com/single-article/industrial-internet-of-things-iiot-benefits-examples/a2fdb5aced1d779991d91ec3066cff40.html. [Accessed: 31- Aug- 2017].

[2] C. Qingping, H. Yan, C. Zhang, Z. Pang, and L. Xu, "A reconfigurable smart sensor interface for industrial WSN in IoT environment." *IEEE Transactions on Industrial Informatics,* vol. 10, no. 2 pp. 1417–1425, 2014.

[3] V. Shnayder, B-R Chen, K. Lorincz, T.T.F. Fulford-Jones, and M. Welsh, "Sensor networks for medical care," Harvard University, Tech. Rep. TR-08-05, Apr. 2005.

[4] D.S. Lee, Y.D. Lee, W.Y. Chung and R. Myllyla, "Vital Sign monitoring system with life emergency event detection using wireless sensor network," In IEEE conference on Sensors, Daegu, Korea, 2006, pp. 518–521.

[5] W. Chung, C. Yau, and K. Shin, "A cell phone based health monitoring system with self-analysis processing using wireless sensor network technology," *In proc. of the international conference of the IEEE EMBS,* 2007, pp. 3705–3708.

[6] S. Koch, M. Hagglund, "Health informatics and the delivery of care to older people," Maturitas 63(3), 2009, pp. 195–199.

[7] O. Omeni, O. Eljamaly, and A. Burdett, "Energy efficient medium access protocol for wireless medical body area sensor networks," 4th IEEE/EMBS International summer school and symposium on medical devices and biosensors, August 2007, pp. 29–32.

[8] G. Manes, G. Collodi, R. Fusco, L. Gelpi, and A. Manes, "A wireless sensor network for precise volatile organic compound monitoring." International Journal of Distributed Sensor Networks 8, no. 4 (2012): 820716.

[9] Y.M. Yuang, M.Y. Hsieh, H.C. Chao, S.H. Hung, and J.H. Park, "Pervasive, secure access to a hierarchical sensor-based healthcare monitoring architecture in wireless heterogeneous networks," IEEE J. Sel. Areas Commun. Vol.27, pp. 400–411, 2009.

[10] G. Zhao, "Wireless sensor networks for industrial process monitoring and control: A survey," Netw. Protoc. Algorithms, vol.3, pp. 46–63, 2011.

[11] G. Manes, G. Collodi, R. Fusco, L. Gelpi, A. Manes, "Continuous remote monitoring in hazardous sites using sensor technologies" International Journal of Distributed Sensor Networks, vol. 8, no. 7, 2012.

[12] K. Lu, Y. Qian, M. Guizani, H.H. Chen, "A framework for a distributed key management scheme in heterogeneous wireless sensor networks," IEEE Trans. Wirel. Commun., vol.7, pp. 639–647, 2008.

[13] X. Du, Y. Xiao, M. Guizani, and H.H. Chen, "An effective key management scheme for heterogeneous sensor networks," Ad Hoc Netw., vol. 5, pp. 24–34, 2007.

[14] P. Traynor, R. Kumar, H. Choi, G. Cao, S. Zhu, T.L. Porta, "Efficient hybrid security mechanisms for heterogeneous sensor networks," IEEE Trans. Mobile Comput., vol.6, pp. 663–677, 2007.

[15] Das ML. Two-factor user authentication in wireless sensor networks. IEEE Transactions on Wireless Communications 2009; 8(3): 1086–1090.

[16] Y. Cheng, and D.P. Agrawal, "An improved key distribution mechanism for large-scale hierarchical wireless sensor networks," Ad Hoc Netw., vol.5, pp. 35–48, 2007.

[17] Le XH, Khalid M, Sankar R, Lee S, "An efficient mutual authentication and access control scheme for wireless sensor networks in healthcare," J Netw., vol. 6, pp. 355–364, 2011.

[18] D.M. Lal, "Two-factor user authentication in wireless sensor networks," IEEE Trans Wirel Commun, vol.8, pp. 1086–1090, 2009.

[19] K. M. Khurram, and K. Alghathbar, "Cryptanalysis and security improvement of 'two-factor user authentication in wireless sensor networks'," Sensors vol.10, pp. 2450–2459, 2010.

[20] H. Debiao, Y. Zhang, and J. Chen, "Cryptanalysis and improvement of an anonymous authentication protocol for wireless access networks," Wirel Pers Commun, vol.74, no.2, pp. 229–243, 2014.

[21] A. Hamed, and M. Nikooghadam., "Three-factor anonymous authentication and key agreement scheme for telecare medicine information systems," J Med Syst., vol.38, no.12, pp. 1–12, 2014.

[22] He D., Kumar N. and Chilamkurti N., "A secure temporal-credential-based mutual authentication and key agreement scheme with pseudo identity for wireless sensor networks," Information Sciences, vol.321, no.10, pp. 263–277, 2015.

[23] F. Al-Turjman, "Impact of user's habits on smartphones' sensors: An overview", IEEE *HONET-ICT*, Kayrenia, Cyprus, 2016, pp. 70–74.

[24] I. Elgedawy, and F. Al-Turjman, "Identity provisioning framework for smart environments", *IEEE HONET-ICT*, Kayrenia, Cyprus, 2016, pp. 12–16.

[25] C. Liqun, Z. Cheng, and N. P. Smart. "Identity-based key agreement protocols from pairings." International Journal of Information Security, vol. 6, no. 4 (2007): 213–241.

[26] M. C. Chuang and M. C. Chen, "An anonymous multi-server authenticated key agreement scheme based on trust computing using smart cards and biometrics," Expert Syst. Appl., vol.41, no.4, pp. 1411–1418, 2014.

[27] Watro R, Kong D, Cuti S, Gardiner C, Lynn C, Kruus P. TinyPK: securing sensor networks with public key technology. Proceedings of the 2nd ACM Workshop on Security of Ad Hoc and Sensor Networks (SASN 2004), Washington, DC, USA, 2004; 59–64

[28] D. Wang, N. Wang, P. Wang, S. Qing, Preserving privacy for free: efficient and provably secure two-factor authentication scheme with user anonymity, Inf. Sci. 321 (2015) 162–178.

[29] Yuan J, Jiang C, Jiang Z. A biometric-based user authentication for wireless sensor networks. Wuhan University Journal of Natural Sciences 2010; 15 (3): 272–276.

[30] E.-J. Yoon and K.-Y. Yoo, "Robust biometrics-based multiserver authentication with key agreement scheme for smart cards on elliptic curve cryptosystem," J Super comput., vol.63, no.1, pp. 235–255, 2013.

[31] Tseng HR, Jan RH, Yangand W. An improved dynamic user authentication scheme for wireless sensor networks. *Proceedings of IEEE Global Communications Conference Exhibition & Industry Forum,* Washington, DC USA, 2007; 986–990.

[32] Lee TH. "Simple dynamic user authentication protocols for wireless sensor networks". *In Proc. of the Int. Conf. on Sensor Technologies and Applications,* Cap Esterel, France, 2008; 657–660.

[33] L. Cheng, W. Li and D. Ma, "Secret Key Generation via Random Beamforming in Stationary Environment", *International Conference on Wireless Communications & Signal Processing (WCSP),* 2015, pp. 1–5.

9
SECURED CACHING IN IoT NETWORKS[1]

Contents

[1] Previously published in F. Al-Turjman, "Fog-based Caching in Software-Defined Information-Centric Networks", *Elsevier Computers & Electrical Engineering Journal*, vol. 69, no. 1, pp. 54–67, 2018.

Abstract

In this paper, we propose a cache replacement approach for Fog applications in Software Defined Networks (SDNs). Our approach depends on three functional factors in SDNs. These three factors are: age of data based on periodic request, popularity of on-demand requests, and the duration for which the sensor node is required to operate in active mode to capture the sensed readings. These factors are considered together to assign a value to the cached data in a software-defined network in order to retain the most valuable information in the cache for longer time. The higher the value, the longer the duration for which the data will be retained in the cache. This replacement strategy provides significant availability for the most valuable and difficult to sense data in the SDNs. Extensive simulations are performed to compare our approach against other dominant cache replacement policies under varying circumstances such as data popularity, cache size, network load, and connectivity degree.

Keywords

Fog computing, Caching, Software Defined Networks.

I. INTRODUCTION

A Software-Defined Network (SDN) is a virtual network capable of acquiring knowledge about its users/inhabitants and its surroundings, and uses such knowledge to help its inhabitants achieve their goals and desires in a context-sensitive manner [1][2]. This definitely improves inhabitants' quality of life, and helps in optimizing and controlling the dramatically increasing consumption rates of resources in smart environments. Inhabitants (users) of a large smart environment (such as a city) could be people, systems, devices, services, or agents, occupied with smart enabling technologies such as RFIDs, sensors, nano-technology, etc. With the evolution of the information-centric IoT the global data networks are interconnected and accessed over cloud systems. The increasing demand for highly scalable and efficient distribution of content/information has motivated the development of future Internet architectures based on named data objects (NDOs), for cxample, web pages, videos, documents, or other pieces of information. The approach of these architectures is commonly called information-centric networking (ICN). In contrast, current networks are host-centric where communication is based on named hosts, for example, web servers, PCs, laptops, mobile handsets, and other devices. Information-Centric Networks serves as a Data-based model which focuses on client's demands disregarding of the data's address or the origin of distribution. ICN is the next generation model for the *Internet* that can cope with the user's requests/inquiries regardless of their data-hosts' locations and/or nature. The current *Internet* model is suffering from the exchange of huge amounts of data while still relying on the very basic network resources and IP-based protocols. Meanwhile, ICNs promise to overcome major communication issues related to the massive amounts of distributed data in the Internet. ICNs adopt a Data-centric architecture which focuses more on the networked data itself rather than the meta-data. This kind of network architectures are known usually by the Content Oriented Networks (CONs) term [3]. Luckily, these CONs architectures match a lot with the emerging communication trend that aims at exchanging Big-data over tiny and energy-limited wireless sensor networks (WSNs) in order to realize numerous attractive projects such as the Smart-planet and the Internet of Things [4][5]. Thus, a new platform is needed to meet these requirements. A new platform, called Fog Computing [6], or, simply, Fog because the fog is a cloud close to the ground, has been proposed to address the aforementioned requirements. Fog is a Mobile Edge Computing (MEC) that puts services and resources of the cloud closer to users to be facilitated in the edge networks.

Unlike Cloud Computing, Fog Computing enables a new breed of light applications and services, that can be run at particular edge networks, such as WSNs. In order to enable WSNs to support this trend in communication and function in a large-scale application platform, such as the Fog Computing, we proposed the cognitive framework from our previous work [7]. In [7], we use smart in-network devices with the capabilities of making decisions based on the information obtained from WSNs to put forward a new information-centric system. The knowledge and reasoning used to dynamically determine the appropriate route where knowledge is defined using value and attribute, and reasoning is represented using the analytic hierarchy process (AHP) technique. The authors in [8] and [9] point out that the upcoming WSN properties such as reliability and delay shall use AHP in their Quality of Information (QoI) assessment. This cognitive Information-Centric Sensor Network (ICSN) framework is able to significantly outperform the *non-cognitive* ICSN paradigms. However, this cognitive ICSN framework did not consider yet the in-network caching feature. Caching in multitude of nodes in ICNs has pivotal role in enhancing the network performance in terms of reliability and response time. In this paper, we propose the use of Value of sensed Information (VoI) cache replacement strategy. It identifies the most suitable data to be replaced in order to maintain prolonged data availability periods while enhancing the network performance. However, authors in [10] and [10] claim that the conventional cache replacement strategy has been intended for IP-based networks and data-centers, which have different data positioning characteristics against the future networks, such as the ICSNs. Moreover, different caching strategies have different effects on the overall performance of the network, and hence, a given caching strategy can influence publishers' load, hit-ratio, and time-to-hit metrics. Numerous attempts in the literature have reviewed each of these metrics independently. However, a single ICSN has the ability to handle multiple users with different designs. Accordingly, a generic dynamic utility function with the ability to consider all the metrics mentioned above while emphasizing on the application itself should be used.

To this end, we provide a novel utility function that sets a value to each cached data item in an ICSN framework. This utility function can determine which data item to drop from the cache while experiencing limited hardware resources for caching. Furthermore, we provide a cache replacement strategy that depends on the VoI in choosing the most appropriate data to be replaced in the cache. We compare our VoI approach against three dominant cache replacement approaches: Node Role-based Caching (RC), Data-based Caching (CC), and Geo-based Caching (GC) with regard to various performance metrics under a variety of parameters including cache size, data popularity, in-network cache ratio, and network connectivity degree.

The rest of this paper is organized as follows. Section II provides an overview of the existing caching approaches in ICSNs. Section III talks about our ICSN-specific system model which we use to build the proposed VoI caching policy. We provide a detailed explanation for the VoI approach in Section IV. Section V presents the detailed simulation results obtained from comparing VoI against other caching approaches. Section VI summarizes our concluding remarks.

II. RELATED WORK

In Fog paradigm, data has to *be close* to the consumers/users. This is the purpose of caching approaches in this paradigm. Caching is associated usually with naming and data delivery approaches/architectures. For instance, the Data-Oriented Network Architecture (DONA) is coupled with naming tuples and labels. Other architectures differ in the basis of retaining data and which entity in the network can keep a copy of the data. Recently, data caching based on how long it was it was in the network is recommended. However, it is quite difficult to claim efficiency when the overhead messages cannot traverse the network. We look at the different Information Centric Sensor Networks caching approaches in this article. Also, we classify the existing caching in ICSNs as follows: A) Geo-based caching, B) Data-based caching, and C) Role-based caching.

A. Geo-Based Caching (GC)

In Geo-based caching, data is cached mainly based on the geographical location of the caching node. For example, Chai et al.

in [12] recommended caching in less spaces in ICSN against caching everywhere. Their policy claims that data should only be cached in nodes with the highest cache-hit rate. Meanwhile, the Cache Aware Target idenTification (CATT) is a topology aware caching policy proposed by Eum et al. [13] where a downloading path is selected given that it has the highest connectivity degree. Nevertheless, this can make this kind of node behave like a geographical bottleneck in the network. Moreover, the authors in [14] have looked into the performance of topology based replica on internet router-level topology and concluded that the router-level fan-out is almost as good as the greedy placement of replica. The node degree used by the work done in [13–15] cannot be considered as a sufficient solution for replica replacement because most of the nodes contain similar, and relatively low degree or fan-out. The author in [14] proposed the use of self-organizing cache management systems, where nodes make globally similar decisions. This system has proven to have reduced delays against the conventional ways and smaller per-node cache. Li et al [21] proposed a selective neighbor caching system, in which a subset of neighboring proxies are selected such that the minimum mobility cost is experienced. This approach is grounded on caching data requests and their corresponding meta-data in a subset of proxies one hop away from the data publisher. Authors in [22] suggests a probabilistic approach for ICNs. They claim that the probability of a file being cached should be increased as it travels from source to destination by considering the following parameters: i) The distance between source and current node, ii) Distance between destination and current node, iii) Time-To-Live for the routed data content, and iv) the Time-Since-Birth. Authors also suggests redundancy in caching on a single path between source and distention. However, this degrades the ICN performance dramatically while experiencing limited caching spaces. Moreover, in [22], authors assume that all the network nodes has the capability of caching, which is not the case in practice with Fog systems. The proposed approach is weak as well due to considering static data request's frequency from a subnet where that data can exist. Nevertheless, we believe caching should be based on dynamic frequencies and location-independent.

B. Data-Based Caching (DC)

Data-based caching is another candidate category for caching ICNs, in which the data replacement decision is taken based on the content of the exchanged data. For example, authors in [16] propose an autonomic cache management architecture that dynamically (re)assigns data items to in-network caches. Distributed managers make (re)placement decisions, based on the observed data request patterns such as their popularity, in order to minimize the overall network traffic. In [17] also authors suggest that every cache manager should decide in a coordinated manner with other cache managers whether or not to cache an item. This approach assumes that every cache manager has a holistic network wide view of all the cache configurations and relevant request patterns. And thus, it adapts depending on the volatility of the user requests. It is evident that the network wide knowledge and cooperation give significant perfor-mance benefits and reduce significantly the time to convergence, but at the cost of additional message exchanges and computational overhead.

In the meantime, other authors are in favor of minimizing the traffic generated by the Internet Service Provider (ISP) and also minimizing the access of in-network devices by caching frequently requested data in the ISP-specific routers. Effective caching is the main problem that is being addressed here, where routers need to organize their data replacement strategy based on their content. The authors have given two data-based caching algorithms. Nonetheless, because the authors have assumed a single gateway in an ISP network, this system may not be practical. Shoa et al. [24] proposed WAVE, which is another data-based caching policy where the size of the cache is adjusted based on data popularity. A node that is in a higher level (called upstream node) suggests the number of chunks of data to be stored in the node at a lower level. This num-ber increases exponentially as the number of requests increases so as to reduce communication and cache management overheard. Additionally, WAVE dispenses data to the network edge. This is where requests come from, putting into consideration the popularity of the data content and the distance relation. Authors in [19] have proposed a data age-based distributed caching system, whose main goal is to reduce the in-network delays and data publisher loads. This system allows for the simple cooper-ative mechanism so as to control where the ages of the data are updated. It distributes the prevalent content to the edges of the ICSN, while at the same time getting rid of the undesirable replicas in-between the ICSN nodes. However, this approach encounters issues from sustaining highly dynamic contents, and hence, nodes that reside far from the server take long periods of time to refresh their data.

C. Role-Based Caching (RC)

In this category, authors consider the role of the in-network caching in order to realize the full capabilities of an ICSN. Additionally, the data that needs to be cached at the control or management level shall be considered. Therefore, authors in [20] have established the effects incurred when handling caching decision at the data level and proposed a new method to deal with caching at the control level. The proposed method can be used to bring a balance between the benefits and cost overhead. Nonetheless, it can only be used in small scales, which means it cannot handle the extensive amount of data found in the internet.

Authors in [23] proposed *LocalGreedy* algorithm for caching in ICNs. They consider a cache cluster consisting of number of leaf nodes which are either directly connected or indirectly via first node as a common parent somewhere along the path to the root node. However, this approach necessitates a global knowledge of the in-network nodes' capability and this contradict with the Fog vision. The authors in [24] discussed the trade-off between caching in a distributed IP-based network and the new emerging systems used in Fog computing, such as the Content–Centric Networks (CCN). They applied their study on real-time traffic generated by functional resources such as the web, file sharing and multimedia streaming. It has been demonstrated that caching videos on the in-network routers increases the number of cache hits. Nonetheless, the other type of data will be better off cached in very large capacity storage area at the core of the network. Hence this type of caching is not efficient in ICSNs.

In recent times however, the *internet* has progressed towards an information centric sensor networking paradigm, where the focus is on delivering named blocks of data to users at the network edge rather than establishing end-to-end connections to the web server. So the design of the cache replacement policy in ICSNs must be a dynamic one, based on the user's request trends, and the application on hand. In this paper, although we need to use a content centric approach, the same cache replacement approaches cannot be applied to an ICSN. This is because of the unique resource constraints of the sensor network, the uncertainty of the wireless medium, and the need to be aware of user-requirements in the ICSN architecture. The resource limitations of the sensor network nodes include limited power supply, storage space, and heterogeneity in terms of the sensors used and the node functions. And this can dramatically affect the route discovery process while locating the cached data. It can significantly degrades the available limited energy in case an unreliable path is considered for data caching [15][23]. In addition, the same content (sensed data) cannot be replicated into multiple caches without associating them with location, because the sensed information may be different in different parts of the network, and it may change over time too, which is unlike the case of ICSNs. This makes the cache replacement trickier in information centric sensor networks. In addition, the replacement policy should take into account the type of user requests coming to the network, the sensor node availability at different locations (as nodes eventually die out), and also the sensing duration for different sensors on board the sensor nodes.

Unlike other attempts, this article proposes dynamic caching decisions by considering knowledge and reasoning elements. Moreover, by considering cognitive observations, we prioritize the cached contents in a hierarchal storage manner. At the same time, we propose a utility function that assigns a data item value based on its specific attributes such as popularity and age. This makes the proposed VoI scheme more appropriate for Fog networks.

III. SYSTEM MODELS

In addition to ICSN's popularity, age and delay models, we also give a detailed explanation on the network model of the SDN-based ICSN in this section.

A. Considered ICSN Model

In our model, there are three main entities: a content provider (i.e., the data publisher/sensor), a service provider (e.g. Internet Service Provider (ISP)), intermediate caching nodes (called relay nodes (RNs) or local cognitive nodes (LCNs)), and a client (i.e., a user/destination for requested data). The elements of cognition found in LCNs such as knowledge, reasoning and learning, are assisting in data requests forwarding. They interact with publishers/sensors, RNs, and the end-user/sink, where all of the collected data are delivered. Data is sent from the publishers/sensors to the sink via RNs as a result of a user request. The sink node is also upgraded with cognitive elements, and hence now on we call it Global Cognitive Node (GCN). The ISP runs a web publishing service by displaying web-based pages for the data publisher, and the user visits the web pages. In general, both the ISP and the publisher will naturally like to encourage the client to reach the nearby web pages on the intermediate caching nodes which contain the targeted data via measuring the web activity [24]. There are many existing enterprises that try to provide services for measuring the activity of web sites. A partial list of these, in [25], includes companies like I/PRO, Nielsen, NetCount, RelevantKnowledge, and others. These companies use mainly two methods: sampling the activities of a group of web clients, and installing an audit module in web sites. Thus, we assume dynamic auditing or assurance services from the publisher and/or ISP side as our system is supposed to help the user finding the targeted data according to a computed value of information (VoI). The assumed system/software autonomy in the proposed approach is clarified based on software agent theory [24] as follows:

1. Client-Based Architecture   The architecture of the client-based side is depicted in Figure 9.1(a). The figure shows the following main modules:

- **Input Proxy:** As our system could be installed on different devices (e.g., sensors, smart phones, tablets, laptops, etc.) that differ in their capabilities and adopted communication protocols.
- **Data AEC:** It is the module responsible for raw data Aggregation, Encryption, and Compression (AEC).
- **Network Connector:** It is the module responsible for sending the AEC raw data blocks to the SDN. So it should synchronize the blocks with the communication proxy located in the publisher side.
- **Output Proxy:** This module is responsible to send the notification and warning messages to the end-user based on the used device capabilities.

(a)

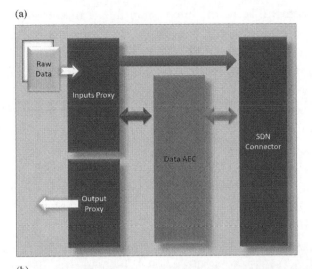

(b)

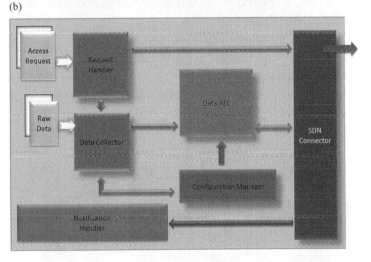

Figure 9.1   (a) Client node architecture, and (b) Caching-node architecture.

2. Caching-Node Architecture The architecture of the caching node is depicted in Figure 9.1(b). The figure shows the following main modules:

- **Request Handler:** It is the module responsible of getting the requests from the client to access a specific data and/or service. The request handler prioritizes the concurrent requests if any and to the most appropriate caching node or publisher in case the data was not cached locally.

- **Data collector:** It is the module responsible of monitoring all the user's interactions, and collecting the corresponding raw data, then sends it to the AEC module to be cached.
- **Configuration Manager:** It is the module responsible for data configuration, in which the caching node specifies which interactions and information to be monitored and collected, also it can specify the preferred cache size, encryption and compression options.
- **Notification Handler:** This module is responsible of collecting the notifications coming from the publisher side to be sent to the end user.

3. Publisher-Based Architecture The architecture of the publisher-based side of our system is depicted in Figure 9.2. It shows the main modules described as follows:

- **SDN Communication Proxy:** the smart device might not be supporting the HTTP protocol (such as Contiki sensors, and other devices), hence this module must perform the mapping process between the mismatching protocols.

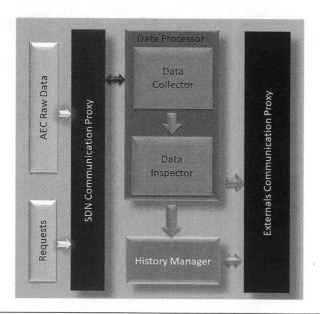

Figure 9.2 Publisher node architecture.

- **Data Processor:** It is the module responsible for collecting the data blocks and requests from the clients, then pass them to the history manager and observer modules. The data processor consists of a data collector and a data inspector. The data collector collects the data from different clients, while the inspector checks if the proper encryption and compression techniques are adopted, also it performs pre-processing steps on the collected data before passing it to the history manager module.
- **History Manager:** It is the module responsible for storing the inhabitants' data blocks and requests, so they can be accessed by the observer module for learning purposes.

Since the users of the information-centric SDNs including people, services, systems, or agents need to interact with each other to create a smart space that improves data availability and access, these interactions need identity verification in such a way that assures security levels. Existing solutions such as static identity approach discussed in [1] and [2] impose a risk of identity theft during the interactions among inhabitants. To maintain a strategic distance from such risks, an innovative approach that uses an Objective-Driven programming has been exposed [26]. The secured objective-driven model can be created automatically at runtime by assuming Provisioning-Assurance-Auditing (PAA) Cloud Engine along with the XACML security annotation representation [26]. Where the later provides a secured separated abstraction layer for the cloud users on top of the programming model. In this research, we assume an encrypted and compressed user profile by using his cached activities' history and usage patterns of the environment's resources. And this allows creating an identity proxy to perform the verification required during the interaction.

B. Delay Model

For different sensors, there is a different amount of time that they should be open to the environment so as to better capture the sensed data. The author in [24] says that, the duration of exposure consequently affects the on-time of the sensor nodes, which in turn affects its life time. When the delay incurred when reading data is more, it is advisable to store the sensed data for long so as to increase the lifetime of the sensor node. This is called sensing delay. Additionally, there will be a propagation delay added if every time data is requested it has to be moved from the sensor node to LCNs. This is most likely to happen, if the sensor node is far from the sink. Hence, the delay components that we consider are sensing delay δ and the propagation delay τ. Furthermore, we limit the number of hops which are needed for data to be delivered at the sink so as to evade the unnecessary usage of energy. Accordingly,

$$\tau \propto n, \quad \text{where } n < 6 \tag{9.1}$$

$$\delta \propto \max(d_1, d_2, d_3, \ldots d_k) \tag{9.2}$$

Where k is equal to the total number of sensing elements found on a sensor node. And d_i is equal to a fixed sensing delay value of the sensor type i. Hence, the sensing delay is a function of the maximum delay. As shown in Equation (9.3), the total delay (Δ) associated with delivering freshly sensed data to the sink is a sum of sensing and propagation delays as follows:

$$\Delta = \tau + \delta \tag{9.3}$$

C. Data-Age Model

The age model uses two approaches to decide which data should be plunged from the cache. First approach is one of the ways which uses periodicity of the periodic request. The second approach is applied when the cache is full. Since newly sensed data has to be given at the beginning of each periodic cycle, we can make use of this periodicity. Hence, when the cache is full at the end of each periodic request, old data from the cache can be flushed out. Hence, the time-to-live (TTL) gives the age of a sensed attribute-value pair. For our model, we don't cogitate on the use of historical data, and hence, cached data can be refreshed at the end of each cycle, provided that the data is transmitted to the sink/GCN at the end of each cycle. Therefore, the cache holding period becomes a function of the transmission cycle's periodicity according to the application needs to hold data for a longer period of time.

D. Popularity Model

A given set of sensed data can be of interest multiple times to a single ore many users, or a number of users can be interested in a given type of sensed data. Such sensed data is considered popular, and hence stored in LCNs' cache for a longer period of time. In order to keep data available when a sensor node start to die out, LCNs should retain data for longer periods of time. When a primary LCN begins to die, it is advisable to store the data in the neighboring LCN, that way they can be stored for longer. This process is managed by a traffic planning algorithm.

E. Communication Model

In this section we talk more on the assumed channel model in our wireless communication. T_{po} Represents the transmission power used by the *ICSN* nodes, while T_r is the transmission range between *BS* and *SN*. The expression of the channel model is given below:

$$C_M = A\rho T_{P_o} T_r^{-\alpha} \qquad (9.4)$$

where C_M is the transmission power of the *BS*, A is a constant gain factor for power from the antenna and amplifier gain, ρ is a small scale constant for the fading factor, and α is the path loss exponent. Accordingly, this model considers not only the available energy per node, but also the surrounding environment conditions.

IV. CACHE REPLACEMENT IN SDNs

In this section, we describe our systematic approach followed in replacing the cache contents in an ICSN. First, all data that take a long time to be collected/found should be stored for longer periods of time so as to conserve more energy on the sensor nodes. Secondly, the storage of data must be a function of the periodicity. This periodicity will assist in keeping the old data version until the new one arrives, and working on requests from different types of traffic in a timely manner. Finally, the value of the data collected can be computed according to a utility function that adapts based on the targeted application.

Hence, it is important to consider how old the data is when working on the requests for data on demand. Therefore, we can apply our cache replacement approach at the LCNs of the network since the criteria is known and fixed. Accordingly, our preposition VoI-based approach uses the abovementioned system models to realize the efficient way in caching while handling the following types of data:

1. Delay-based data: cached data has a delay sensitivity which is a parameter specified by the user to show how long they are willing to wait for the data. This delay-sensitive data can be found in areas which require emergencies such as disaster or health emergencies.
2. Demand-based content: This is how popular a set of data can be, which is obtained by how often the data is requested.
3. Age-based content: Some data are sensitive to time, for instance, if a user requires information about the city traffic for the next one hour, any information outside this time limit has no use.

Consequently, VoI cache management scheme used three factors to set VoI_{Si} for each sensor node S_i reading. This value depends on the content in each operational round. The function below is used at the beginning of every round as mentioned before to reset VoI_{Si}.

$$VoI_{Si} = \alpha * \Delta + \beta * TTL_{Si} + \gamma * Popularity_{Si} \qquad (9.5)$$

where α, β, and γ are the factors that are itemized based on the user request and the type of traffic. The ability of VoI to get priorities based on ICSN gives it lead. Hence, for better priority caching, we must adjust some parameters to obtain an effective approach. We use the delay sensitivity parameters to minimize delay. The popularity factor is critical because it tells us which set of data has had the most frequent requests. The packet age factor is also important in that it lets us know which data has not been used for a long and replaces it with relevant data. The following algorithm provides steps to be followed by each node, if its cache is full so as to expel data with the minimum VoI_{Si}.

Moreover, VoI can be based on the combined value of the abovementioned QoI attributes (e.g., delay and popularity), and energy consumed during the process of delivering information to the GCN. VoI delivered to the end user is said to be maximized when data is delivered over links that provide the best effective QoI for each traffic type, while minimizing the energy consumed in the network while doing so.

$$VoI = \sum_{n-hops}\left(Effective\ QoI\right) - \sum_{n-hops}\left(Energy\ Cost\right) \qquad (9.6)$$

Equation (9.6) highlights that the lower the energy cost in delivering data to the sink, the higher the VoI associated with that data/information object. The QoI must be maximized and energy cost minimized to achieve the best VoI value. If energy consumption is measured as a function of the number of transactions taking place before data is delivered to the GCN, a simple metric - the hop count can be used to approximate the energy cost. If the information is transmitted from source to GCN over minimum number of hops, each link providing the best combined QoI for that traffic type, we can say that the information was delivered to the GCN with good VoI.

Algorithm 1: Drop least VoI_{Si}.

1. **Function VoI** (*content*)
2. Input
3. *content: A content item within the SDN.*
4. **Begin**
5. **for** each LCN node, **do**
6. **for** each duty cycle, **do**
7. **Set** *value* of each VoI_{Si} in the cache based on Eq. (5)
8. **if** *cache_full*
9. Check history of user requests
10. Drop the data content of the least VoI_{Si}
11. **End if**
12. **End for**
13. **End for**
14. **End**

In the above Algorithm, elements of cognition are implemented at the LCNs. These elements are: *Learning and Reasoning* elements.

A. Learning

Learning is used in our VoI approach in order to determine the most appropriate paths towards the GCN that satisfy the Fog network requirements. This cognition element uses a direction-based heuristic to determine the data delivery path through RNs that lie in the direction of the GCN. Hence, each time a LCN has to choose the next hop, the direction-based heuristic eliminates RNs that increase the distance between the current RN and GCN. This information is stored in the LCN for use in the next transmission rounds. Thus the direction-based heuristic, along with feedback from the network about the chosen paths helps the LCNs to learn data delivery paths to the sink, as the network topology changes.

Example 1: Assume S_1 and S_2 have data to be sent to destination nodes D_1 and D_2. R_n are all the available relays towards the destination. Out of these relays, it is determined that R_5 as shown in Figure 9.3(a) has the lowest link outage probability to D_1 and D_2. Therefore, S_1 initiates routing data to R_5. Meanwhile, S_2 also forward a high traffic of data to R_5 (depicted by solid paths in Figure 9.3). When multiple source nodes start routing their data to R_5 as well, the route to R_5 may get congested. A cognitive network with *learning* capabilities will be able to identify the congestion at R_5 (by observing the decrease in throughput). Sharing this observation with neighboring nodes, the cognitive Fog network (or ICSN) would be able to respond to the congestion proactively, by routing the data through a different path involving nodes R_4, R_8 and R_9 as shown in Figure 9.3(b).

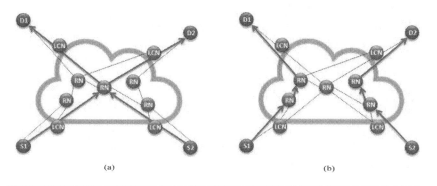

(a)   (b)

Figure 9.3 (a) Classical routing and (b) Cognitive routing in Fog: (a) Classical case in a sensor network, (b) Cognitive case in response to congestion.

B. Reasoning

In the VoI approach, we assume a modified version of the Analytic Hierarchy Process (AHP) [7] for implementing the reasoning element of cognition in the Fog network. AHP supports multiple-criteria decision making while choosing the data path. For example, if we have a delay-sensitive data, the node which provides the lowest latency, will be chosen even though it might degrade other metrics such as the network energy or throughput. If two next-hops guarantee the same latency then the next attribute to compare will be energy, and then, throughput, assuming that energy is the next desired attribute in the Fog network. AHP provides a method for pair-wise comparison of each of the attributes and helps to choose the node that can provide the best network performance on the long run. The following example has more details on the utilized AHP.

The steps used in the AHP to establish priorities for the QoI attributes and identify the best next-hop path in delivering the application data to the GCN are illustrated in Algorithm 2.

Algorithm 2: AHP analysis to determine the data delivery path.

1. Function AHP (QoI.priorities)
2. **Input**
3. QoI.priorities: End-user defined priorities on QoI attributes for requested data
4. **Output**
5. RN_x: A selected best next-hop $RN_x \in \{ RN_1 \ ..... \ RN_n \}$ to deliver data
6. **Begin**
7. **Initialize**: *Maxhopstosink*=k; *Expected QoI priorities on {L, R, T}; currenthop=1; flag=0*
8. **While** (*currenthop < Maxhopstosink*)
9. *AHP_analysis*(Next-hop LCNs v/s QoI attributes)
10. Next hop LCN = LCN_x //This is the LCN with QoI priority ratio consistent with expected values
11. Transmit and aggregate data at next-hop LCN
12. **If** (next hop = GCN)
13. **If** ((actual QoI < (.5) *expected QoI) for all attributes)
14. **If** (flag+1 <5)

15. Display: "Poor network performance";
16. **Else**
17. Display: "Network End of Life";
18. Exit;
19. **End**
20. **Else,**
21. Display: "QoI requirement met"
22. Exit;
23. **End**
24. **Else**
25. *currenthop= currenthop+1*
26. goto step 8
27. **End**
28. **If** (*currenthop > Maxhopstosink*),
29. GCN Retransmits request
30. Flag condition to learning algorithm in LCNs
31. **End**

Information about the relative priorities of the QoI attributes as desired by the user are received as input from GCN in steps 1–3. The output is a next hop RN that provides the best QoI as shown by steps 4–5. A maximum value is set for the hop count, within which data is expected to reach the GCN from its source. In step 9, AHP analysis identifies the best next-hop RN that satisfies these requirements. Actual values of latency, reliability and popularity are used during AHP analysis. The priority values are obtained by calculating the Eigen vector of the matrix and normalizing it. Thus, priorities for each QoI attribute that the RNs offer are obtained. Steps 13–19 help to identify conditions in which the network might be performing poorly, or is not able to deliver data with the expected QoI, in which case it considered as network's end-of-life as it is not providing useful information to the user. Steps 8–26 are iteratively run through till the GCN is reached. Steps 28–31 indicate specific restrictions on the data delivery path.

V. PERFORMANCE EVALUATION

In this section, we provide initial performance evaluation results for the VoI based cache replacement technique, which we have compared with the Role-based Caching (RC), Data-based Caching (DC), and Geo-based Caching (GC) techniques using NS3, a discrete event simulator. The caching schemes: RC, DC, GC, and VoI, are executed on 500 randomly generated wireless heterogeneous network topologies in order to get statistically stable results. The average results hold confidence intervals of no more than 5% of the average values at a 95% confidence level. We make use of the Cache Hit Ratio to compare the performance of the different cache replacement strategies. Cache hit ratio is defined as the ratio of the number of times requested data was found in the cache divided by the total number of times data was requested from the cache. The storage cache is implemented as a single storage level in one case (L1 cache) and as a hierarchy of two storage levels in another case (L1 and L2 cache). Simulation results are compared for VoI, GC, DC and RC replacement techniques. These simulations were run at a cache sizes ranging from 10 to 100, and the simulations end after serving 1000 packet requests. There are 100 different requests from which the packet requests are randomly generated.

A. Performance Metrics

To compare the performance of the proposed VoI approach, we track ICSN-specific metrics to achieve qualitative conclusions for the targeted in-network caching problem. We simulate the performance of an ICSN network with the detailed physical layer NS3 built-in parameters so that we achieve realistic simulation instances. The four considered performance metrics are as follows:

- Cache-hit ratio: is simply the fraction of time a request arrives at a node to which that cache is attached but does not contain the requested data item. It is the average hitting ratio over all the in-network caches.
- Time-To-Hit-data (TTH): is found by simply logging all the total costs of the request and response paths incurred by every sensor node.
- In-network latency (delay): this metric represents the end-to-end delay as described above. Note that we differentiate between latency to hit data and in-network latency since the

two metrics may differ because of mobility or disruption conditions.

- Average Request per Publisher (ARP): this metric is measured in number of data request per hour (req/hr) and it represents the average load per publisher in an ICSN paradigm.

B. Simulation Parameters

Many of the ICSN paradigm parameters have to remain fixed while our simulation instances are generated. In particular, the parameters of our simulation are as follow:

- Percentage of nodes with caches (PoC): This parameter is our primary method for controlling the extent of caching in our ICSN. By varying this parameter, we can study the sensitivity of metrics like time-to-hit-data to the caching extent.
- Connectivity level (degree): It represents how much tightly connected is the ICSN network. We use the connectivity matrix, based on our described communication model in Section III.
- Data Popularity: It indicates how frequent a specific data content is requested. This metric is measured in percentage with respect to other requested data contents.

C. Simulation and Results

The following figures depict the achieved results. Our first objective is to confirm that increasing the extent of caching in ICSNs, in terms of both size and number of levels will reduce time to meet data for all cache policies.

According to Figure 9.4, we can deduce that the VoI is not efficient in level one cache, however, RC and GC cache replacement techniques perform equally well in this scenario. In Figure 9.5, where we have two levels of caching (one on LCN and the other on RN) we find out that VoI surpass the other two replacement techniques. Since we have only one type of the considered requests, there is minimum performance gain when the cache size is increased beyond 30 *Mbyet*.

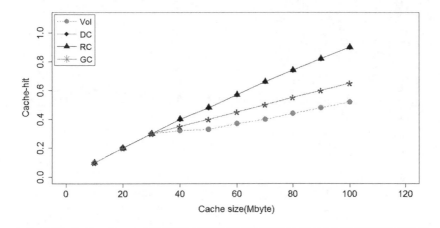

Figure 9.4 Cache size vs. the hit ratio with 1-level caching.

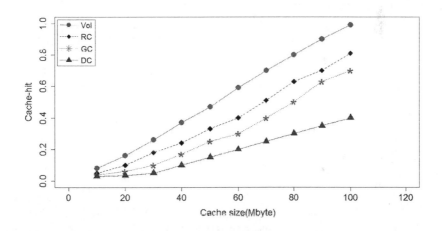

Figure 9.5 Cache size vs. the hit ratio with 2-level caching.

The next set of simulations' figures (Figures 9.6 and 9.7) are set to analyze the performance of the cache replacement strategies as the number of requests that a given network needs to serve increases from 500 to 5000 *req/hr*. The cache size is set to be 100 *Mbyte* and the number of request types are fixed at 4.

From the Figures 9.6 and 9.7, we can deduce that the advantage held by Value of Information (VoI) replacement technique is that it replaces data based on user requirement and VoI of the data. Other replacement techniques are only concerned with the match of the data

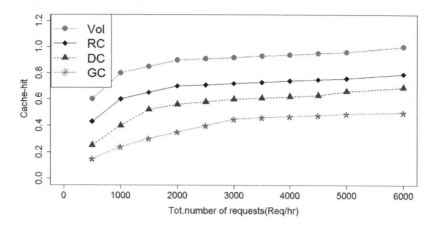

Figure 9.6 Total no. of requests vs. the hit ratio with 1-level caching.

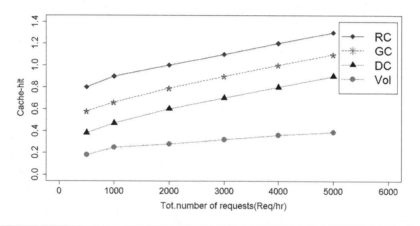

Figure 9.7 Total no. of requests vs. the hit ratio with 2-level caching.

requested packet number, without considering the age of the data, its popularity or the delay associated with sensing and transmitting it to the sink. Nevertheless, the use of VoI replacement technique puts into consideration the age of data, VoI and the popularity of the information. New information replaces old ones, unlike other technique that only find a number match irrespective of their age. Based on this we proposed the use of two level cache, one on LCN and the other on RN, in order to reduce the complexity of computation, we can employ the use of VoI base replacement technique on LCN and RC or GC on

Table 9.1 Two-Level Caching Comparison

L1 CACHING	L1 HIT RATIO	L2 CACHING	L2 HIT RATIO	CUMULATIVE HIT RATIO
Vol	0.900	**Vol**	0.009	0.817
Vol	0.811	**GC**	0.813	0.825
Vol	0.647	**RC**	0.779	0.610
GC	0.799	**GC**	0.500	0.751
GC	0.733	**Vol**	0.402	0.683
GC	0.900	**RC**	0.000	0.718
RC	0.897	**RC**	0.000	0.651
RC	0.900	**GC**	0.000	0.718
RC	0.899	**Vol**	0.000	0.618

RN. We can set the size of level 1 and 2 to 100 and the packet request can be 10000. We can set the size of level 1 and 2 to 100 *Mbyte* and number of packet requests is set to be 10000 (Table 9.1).

From the above table, we can see that for the two levels of cache, the best possible combinations are: VoI based replacement strategy at L1 cache and VoI or GC based replacement strategy at L2 cache. Despite having a good hit ratio sometimes, RC is not reliable to serve the user needs considering the age of data and the delay required to retrieve the required data, because the decision making is only at LCNs. Implementing the VoI replacement technique at LCNs can greatly help save resources especially when a cache hit is found on the first level of the cache.

Figures 9.8 and 9.9 below represents the findings from the simulation experiment of the 2 level cache. From both figures, the extend of

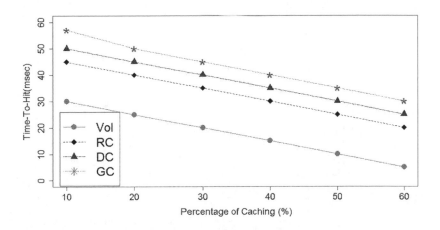

Figure 9.8 Time to hit ratio vs. percentage of nodes with caches (conn. degree = 30).

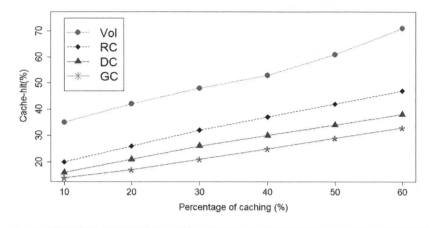

Figure 9.9 Hit ratio vs. percentage of nodes with caches (connectivity degree = 30).

cache availability increases proportionally. According to Figure 9.8, we observe that the overall time to meet data, which is our main performance metric, is reduced in all performance policies. However, the Vol policy performs best at higher proportions of nodes attached to caches. On the contrary, Figure 9.9 shows that there is an increase in the data hit for all the approaches, and hence, we conclude that the Vol is better due to its ability to replace the most relevant data according to an ICSN-specific set of attributes.

In Figures 9.10, 9.11, and 9.12, connectivity level (degree), is the examined parameter. From Figure 9.10, we deduce that there is an increase in

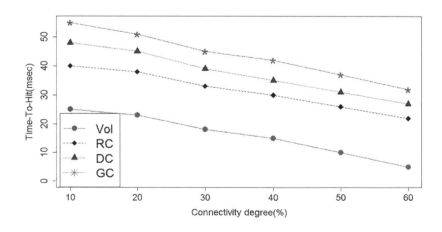

Figure 9.10 Time to hit ratio vs. the connectivity degree percentage.

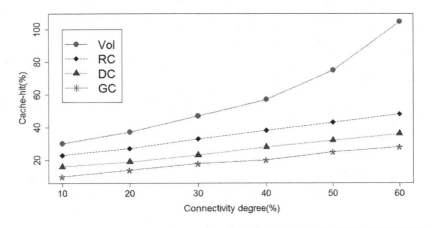

Figure 9.11 Hit ratio vs. the connectivity degree percentage.

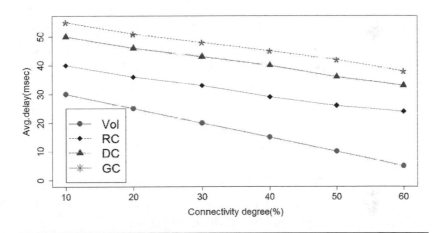

Figure 9.12 Avg. in network delay vs. the connectivity degree percentage.

time to hit data as the ICSN connectivity increases in all the approaches. However, we notice that VoI is less dependent on the network and hence better than the other two approaches. VoI is more dependent on data type- Highly desired property in ICSN network. Figure 9.11 shows the data hit performance against a varying network connectivity degree, and while applying the VoI scheme, we notice that the data hit increases exponentially while the network connectivity increases. Nevertheless, the data hit of the other two approaches increases linearly. Moreover, Figure 9.12 shows that VoI is the best in terms of delay. This can be attributed to the application of the delay factor while deciding which

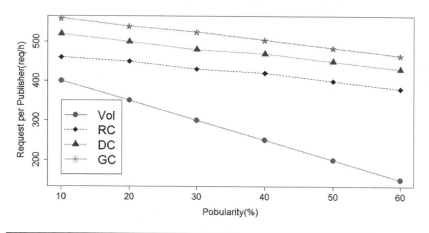

Figure 9.13   Publisher load vs. the data popularity.

data to replace. Figure 9.13 shows the effect of data popularity in terms of publisher load. The VoI tops GC and RC as the popularity metric increases. This is a very desirable property in the ICSNs.

The VoI based technique would be suitable for the ICSN approach, as we propose to use a named data association for the sensed data, such as attribute-value pairs, and the cache size can be decided based on the different types of user requests that the network is expecting to serve. We can expect that the users are more satisfied with the response received from the LCN, as it retains information in its cache based on both data popularity and various parameters that affect gathering sensed information and the energy involved in doing so, as the network scales up to larger sizes.

VI. CONCLUSIONS & FUTURE WORK

To conclude, we note that the VoI-based approach is well-suited in ICSNs while using the name-data association and supporting node mobility. We presume that users will be contented with the feedback provided by ICSNs since they keep information based on data popularity and other parameters that affect the availability of sensed information and energy consumption. Moreover, the fact that data can still be provided from local cognitive nodes (LCNs) even after the node has been died, proves that VoI cache replacement technique assists in a smooth dilapidation of the network.

Additionally, we looked into the influence of changing network loads and the inter-LCN communication on how effective the cache replacement approach is in Fog networks. Subsequently, VoI has a delay-tolerant requirement at the edge of the network, and this provides a dynamic data replacement. By plummeting the load of the data publishers, VoI can maximize their gain as well. The consumer gain is maximized in reference to metrics such as delay, time-to-live, and data popularity.

One of the aspects of the information-centric capabilities, is its ability to cache information in cognitive nodes and use it collaboratively for information sharing across the network. While we acknowledged this advantage in this article, such caching would rise an issue with the mobility-enabled node presence, which we did not delve deeply into yet. Thus, exploring the role of caching in information access and data delivery, and the study of cache replacement techniques that suit the cognitive nodes in mobile ICSNs is still a direction to explore out of this work. Moreover, the idea of cognition can be investigated more while being applied to intermediate relays/switches of the current Internet infrastructure to realize the cognitive network concept in general.

REFERENCES

[1] K. Lal, A. Kumar, ICN-WiMAX: An Application of Network coding based Centrality-measures caching over IEEE 802.16, Procedia Computer Science, vol. 125, 2018, pp. 241–247.

[2] F. Al-Turjman, "5G-enabled Devices and Smart-Spaces in Social-IoT: An Overview", *Elsevier Future Generation Computer Systems*, 2017. DOI: 10.1016/j.future.2017.11.035.

[3] I. Amo, J. Erkoyuncu, R. Roy, S. Wilding, "Augmented Reality in Maintenance: An information-centred design framework", Procedia Manufacturing, vol. 19, 2018, pp. 148–155.

[4] SmartSantander, Future Internet Research and Experimentation. [Online]. Available:http://www.smartsantander.eu/.

[5] F. Al-Turjman, "Information-centric sensor networks for cognitive IoT: an overview", *Annals of Telecommunications*, vol. 72, no. 1, pp. 3–18, 2017.

[6] S. Chen, T. Zhang, W. Shi, "Fog computing", *IEEE Internet Computing*, vol. 21, no. 2, pp. 4–6, 2017.

[7] G.T. Singh and F.M. Al-Turjman, "A Data Delivery Framework for Cognitive Information-Centric Sensor Networks in Smart Outdoor Monitoring", *Elsevier Computer Communications*, vol. 74, no. 1, pp. 38–51, 2016.

[8] M. Z. Hasan, and F. Al-Turjman, "Evaluation of a Duty-Cycled Asynchronous X-MAC Protocol for Vehicular Sensor Networks", *EURASIP Journal on Wireless Communications and Networking*, 2017. DOI: 10.1186/s13638-017-0882-7.

[9] F. Al-Turjman, "Mobile Couriers' Selection for the Smart-grid in Smart cities' Pervasive Sensing", Elsevier Future Generation Computer Systems, vol. 82, no. 1, pp. 327–341, 2017.

[10] M. Chen; Y. Qian; Y. Hao; Y. Li; J. Song, "Data-Driven Computing and Caching in 5G Networks: Architecture and Delay Analysis", *IEEE Wireless Communications*, vol. 25, no. 1, pp. 70–75, 2018.

[11] X. Yue; Y. Liu; J. Wang; H. Song; H. Cao, "Software Defined Radio and Wireless Acoustic Networking for Amateur Drone Surveillance", *IEEE Communications Magazine*, vol. 56, no. 4, pp. 90–97, 2018.

[12] W. K. Chai, D. He, I. Psaras and G. Pavlou "Cache "Less for More" in Information-Centric Networks (Extended Version)", *Elsevier Computer Communications*, vol. 36, no. 7, pp. 758–770, 2013.

[13] S. Eum, K. Nakauchi, Y. Shoji, N. Nishinaga, M. Murata, "CATT: Cache aware target identification for ICN", *IEEE Communications Magazine*, vol. 50, no.12, pp. 60–67, 2012.

[14] M. Meddeb; A. Dhraief; A. Belghith; T. Monteil; K. Drira, "How to Cache in ICN-Based IoT Environments?", *In Proc. Of the IEEE/ACS 14th International Conference on Computer Systems and Applications (AICCSA)*, Hammamet, Tunisia, 2017, pp. 1–6.

[15] F. Al-Turjman, "Energy–aware Data Delivery Framework for Safety-Oriented Mobile IoT", *IEEE Sensors Journal*, 2017. DOI: 10.1109/JSEN.2017.2761396.

[16] B. Nguyen, H. Phan, D. Ha, G. Nguyen, "An Information-centric Approach for Slice Monitoring from Edge Devices to Clouds", *Procedia Computer Science*, vol. 130, 2018, pp. 326–335.

[17] F. Al-Turjman, "Optimized Hexagon-based Deployment for Large-Scale Ubiquitous Sensor Networks", *Springer's Journal of Network and Systems Management*, vol. 26, no. 2, pp. 255–283, 2018.

[18] L. Gkatzikis, V. Sourlas, C. Fischione, I. Koutsopoulos, "Low complexity content replication through clustering in Content-Delivery Networks", *Computer Networks*, vol. 121, no. 5, 2017, pp. 137–151.

[19] Z. Ming; M. Xu; D. Wang "Age-based Cooperative Caching in Information-Centric Networks", *Int. Conf. on Computer Communication and Networks (ICCCN)*, 2014.

[20] W. Yaogong, K. Lee, B. Venkataraman, et al. "Advertising Cached Contents in the Control Plane: Necessity and Feasibility", *In Proc. INFOCOM Workshop on computer communications*, 2014.

[21] H. Li; H. Zhou; W. Quan; B. Feng; H. Zhang; S. Yu, "HCaching: High-Speed Caching for Information-Centric Networking", *In Proc. of the IEEE Global Communications Conference (GLOBECOM)*, Singapore, Singapore, Dec. 2017, pp. 1–6.

[22] A. Gupta, U. Shanker, SPMC-CRP:A Cache Replacement Policy for Location Dependent Data in Mobile Environment, Procedia Computer Science, vol. 125, 2018, pp. 632–639.

[23] T. Baker, B. Al-Dawsari, H. Tawfik, D. Reid, "GreeDi: An energy efficient routing algorithm for big data on cloud", *Ad Hoc Networks*, vol. 35, pp. 83–96, 2015.

[24] C. Zuo, J. Shao, G. Wei, M. Xie, M. Ji, CCA-secure ABE with outsourced decryption for fog computing", *Future Generation Computer Systems*, vol. 78, no. 2, 2018, pp. 730–738.

[25] R. Caropreso; R. Fernandes; D. Osorio; I. Silva, "An Open Source Framework for Smart Meters: Data Communication and Security Traffic Analysis", *IEEE Transactions on Industrial Electronics*, 2018. DOI: 10.1109/TIE.2018.2808927.

[26] M. Gomes, M. Pardal, "Cloud vs Fog: assessment of alternative deployments for a latency-sensitive IoT application", Procedia Computer Science, vol. 130, 2018, pp. 488–495.

10

SEAMLESS KEY AGREEMENT FOR PUBLIC SAFETY NETWORKS[1]

Contents

[1] This work was supported by a Newton Fund Institutional Links grant, ID 216429427. The grant is funded by the UK Department of Business, Energy and Industrial Strategy (BEIS) and managed by the British Council.

Fadi Al-Turjman is with Middle East Technical University, Northern Cyprus Campus, Kalkanlı, Güzelyurt, Mersin 10, Turkey (e-mail: fadi@metu.edu.tr).

Yoney Kirsal Ever is with Near East University, Nicosia, North Cyprus (e-mail: yoneykirsal.ever@neu.edu.tr).

Enver Ever is with Middle East Technical University, Northern Cyprus Campus, Kalkanlı, Guzelyurt, Mersin 10, Turkey (e-mail: eever@metu.edu.tr).

Huan Nguyen is with Middlesex University, London, UK (e-mail: h.nguyen@mdx.ac.uk).

Deebak Bakkiam David is with Middle East Technical University, Northern Cyprus Campus, Kalkanlı, Guzelyurt, Mersin 10, Turkey (e-mail: HYPERLINK "mailto:deebak@metu.edu.tr" deebak@metu.edu.tr).

Previously published in F. Al-Turjman, Y. K. Ever, E. Ever, H. Nguyen, D. Deebak, "Seamless Key Agreement Framework for Mobile-Sink in IoT based Cloud-centric Secure Public Safety Networks", *IEEE Access*, vol. 5, no. 1, pp. 24617-24631, 2017.

Abstract

Recently, Internet of Things (IoT) has emerged as a significant advancement for Internet and mobile networks with various public safety network applications. An important use of IoT based solutions is its application in post-disaster management, where the traditional telecommunication systems may be either completely or partially damaged.

Since enabling technologies have restricted authentication privileges for mobile users, in this study a strategy of mobile-sink is introduced for the extension of user authentication over cloud-based environments. A seamless secure authentication and key agreement (S-SAKA) approach using bilinear pairing and elliptic-curve cryptosystems is presented. It is shown that proposed S-SAKA approach satisfies the security properties, and as well as being resilient to node-capture attacks, it also resists significant numbers of other well-known potential attacks related with data confidentiality, mutual authentication, session-key agreement, user anonymity, password guessing, and key impersonation. Moreover, the proposed approach can provide a seamless connectivity through authentication over wireless sensor networks (WSNs) to alleviate the computation and communication cost constraints in the system. In addition, using - BAN logic, it is demonstrated that the proposed S-SAKA framework offers proper mutual authentication and session key agreement between the mobile-sink and the base-station.

Keywords

Secure public safety networks, Internet of Things, cloud systems, session-key agreement, bilinear pairing.

I. INTRODUCTION

The Internet of Things (IoT) is a novel paradigm where objects become part of the Internet. It has converged technologies in terms of sensing, computing, information processing, networking and controlling intelligent technologies [1, 2]. Among the technologies converged, we can count wireless sensor networks (WSNs), intelligent sensing, remote sensing, radio frequency identification (RFID), near field communications (NFC), low-energy wireless communications, and cloud computing. The technologies involved have particular applications in public safety as well as other domains such as health monitoring, smart homes and environments, smart cities, smart grid, and various types of pervasive systems [3].

WSNs are composed of base-stations and numerous low cost mobility nodes which have restricted resources, such as communication, storage and computation cost. Each mobility node has its own sensing-unit, data-processing unit, module for short-range communication and power-supply unit [9]. Recently, WSNs have had its own prominence in various application fields, namely military (missile target tracking / detection system), environment (hazardous detection), biomedical (health monitoring and patient tracking) and building (smart-homing and threat detection). Since WSNs have limited power-supply unit for the mobility nodes, some researchers [7, 8] have introduced the technique of mobile-sink in the WSNs for the extension of network lifetime. Since the mobility nodes transmit the confidential data via wireless channels, any user may act as an adversary to overhear / tamper the confidential data being transmitted on the WSNs. Lately, Cloud Computing (CC) techniques have been further emerged as well with the WSNs' for the purposes of storage and data access at any time over the Internet [3, 4]. In the cloud, the user can find the set of hardware devices, network connections, storage spaces, data services and application interfaces that are easily accessed over the Internet.

IoT architecture can be implemented as either Internet centric or object centric. The former aims at provisioning services within the Internet, where data are contributed by objects and vendors who deterministically deploy these objects, whereas the latter aims at provisioning services via network of smart objects. Scalability and cost efficiency of IoT services can be achieved by the integration of cloud-computing into the IoT architecture, i.e., cloud-centric IoT [50]. In a cloud-centric IoT framework, sensors provide their sensed data to a storage cloud as a service, which then undergoes data analytics and data mining tools for information retrieval and knowledge discovery [50].

With the evolution of IoT, the global data networks are interconnected and accessed over CC networking systems [4, 5]. As the sensing data are transmitted over public networks, the adversary can easily intercept the exchange of data between the users and the remote servers. This would cause various possible attacks, such as replay, key impersonation, stolen verifier, etc. [14,15]. As the CC has become a prominent domain for secure authentication, WSNs are in high demands of security schemes for the purpose of user authentication, authorization and accounting while the cloud services are being accessed by the legitimate users.

In literature, the gateway/base-station based authentication schemes have provided the lightweight authentication for the enrichment of security properties [10, 11, 12, 13, 14, 15, 16, 17, 18, 19, 20, 21, 22]. As a result, WSNs presume the gateway/base-station as imperative part to sense the real-time data over insecure networks. In this paper, users are referred as mobile-sink. In a state of proper access towards the sensor-node, the mobile-sink should complete the proper registration to establish an authorized session between the sensor-node and the base-station. In addition, the successful establishment of this communication can only be achieved through the satisfaction of mutual authentication and session-key agreement. Generally, the authentication schemes try to satisfy all the security properties of authentication and key protocol (AKA), such as mutual authentication, session key agreement, user anonymity, etc [14]. The analyses performed on the existing studies show that several authentication schemes are still susceptible for various potential attacks, such as privileged-insider, key impersonation, stolen smart-card etc. As a result, this paper presents a seamless secure authentication and key agreement (S-SAKA) using bilinear pairing and elliptic-curve cryptosystems. The objective of this framework is to provide mutual authentication, session-key agreement, data confidentiality, user anonymity, intractability and resilient to node-capture attack, key impersonation, replay, stolen smart card, and privileged-insider. In order to ease the computation and communication overhead, the authentication phase of S-SAKA does not invoke the base-station to authenticate the mobile-sink and sensor-node; and thus the S-SAKA framework is more flexible than the three-party authentication scheme when mobile-sink is employed in the Cloud WSNs.

Since the pairwise keys are randomly distributed, the adversary may have a chance to obtain a common session-key to compromise the nodes. Das et al. [23] and He [24] presented a dynamic-identity based authentication scheme to resist the attacks, like privileged-insider and key-compromise. However, Das et al. and He's approaches are still vulnerable to the potential attack of node-capture. For the enhancement of security efficiency, Deebak [17], Turkanovic et al. [18], Farash et al. [19], Das et al. [20], Amin et al. [21] and Srinivas et al. [22] have proposed lightweight user authentication schemes. However, their authentication schemes fail to mitigate the computation and communication efficiency of the network systems as they invoke the base-station authentication. This paper proposes the S-SAKA framework, which does not only to improve data security while using the mobile-sink in the WSNs, but also provides seamless connectivity over WSNs to reduce the computation and communication overhead.

The major contributions of S-SAKA framework are as follows:

1. The existing authentication schemes [17, 18, 19, 20, 21, 22] are thoroughly analyzed to show various susceptibilities, such as privileged-insider, key impersonation, denial of service and password guessing.
2. To address the security weaknesses of existing schemes [17, 18, 19, 20, 21, 22, 23, 24], a lightweight S-SAKA framework is proposed that holds all the original merits of the existing schemes [17, 18, 19, 20, 21, 22] to resist the potential attacks.
3. To strengthen the proposed S-SAKA framework, a formal security analysis is performed using BAN logic [25]. Besides, the informal analysis is presented to claim that the proposed S-SAKA framework can be resilient to the attacks, which has not been analyzed in the literature to date.
4. Lastly, an experimental analysis is performed using MIRACLE C/C++ library to examine the computation and

communication overhead of existing and proposed authentication frameworks. The evaluation result proves that the proposed S-SAKA framework provides less overhead as compared to existing authentication schemes.

When detailed analysis is carried out with formal and informal verifications, it is observed that the proposed S-SAKA scheme provides less communication overhead in comparison with other existing authentication schemes in the literature [17, 18, 19, 20, 21, 22].

The rest of the paper is organized as follows. Section II discusses the existing secure authentication schemes. Section III illustrates an architecture of hierarchical WSNs and discourses the mathematical assumption model using bilinear pairing. Section IV presents seamless secure authentication and key agreement (S-SAKA) framework along with the security analysis. Section V shows the verification proof. Section VI compares the performance efficiency of proposed and existing authentication schemes. Finally, Section VII concludes this study.

II. RELATED WORKS

Various natural or man-made disasters such as earthquakes, floods, tsunamis, nuclear power plant explosions cost significantly in terms of assets/infrastructure damage and more importantly human lives. The WSN based systems such as IoT solutions can help us to save lives since healthy communication and accurate information can make a real difference between life and death for those who are in the areas affected by the disasters. The exposure of sensitive information or similar attacks on confidentiality/integrity of information, and/or availability of resources can become an additional disaster in case proper countermeasures are not planned carefully.

With the modernization of the public safety communications, and the changes in application areas as well as new technologies introduced such as wearables, wireless body area networks, and variety of tracking devices that can be carried by responders such as rescue teams, fire fighters, and police, the IoT is expected to form a solid infrastructure for public safety applications [4]. Furthermore, although the enhancements especially in performance improvements of 3GPP LTE-A look very promising, during disaster situations these infrastructures can also be damaged or out of service [5]. There are some studies focussing on secure wireless powered device-to-device (D2D) communication in case the infrastructure is not available or partially functional [6]. However IoT based public safety networks (PSNs) are expected to have better availability in disaster scenarios since the computation is known to be more towards the distributed fashion.

Nowadays, sensor nodes are mostly used to sense the continuous data, event detection in real time environment and actuators control. These features are particularly useful for public safety applications. Specifically, micro sensing and seamless wireless connectivity became the promising technologies for various information and communication domains. These technologies are further extending for the classical categories, such as bio chemical processing, space exploration and disaster environment [9]. In order to offer better services to the users in WSNs, security is an important concern as the data transmission is performed over public networks [10, 11, 12, 13] with the restrictions as follows:

1. Sensors are easily render to failure
2. Topologies of sensor networks change often
3. Sensor networks always prefer broadcast paradigms, but most of the Ad-hoc networks are point-to-point communication
4. Sensors have limited power, computation and storage

WSNs are one of the essential components of the infrastructures employed for establishment of IoT based public safety applications. Recently, security issues in WSNs have gained much attention of the researchers not only to satisfy the security properties of authentication and key agreement (AKA) protocol but also to mitigate the computation and communication cost of the system. For the achievement of minimum overhead, several lightweight authentication schemes have been proposed [26, 27, 28, 29, 30]. Watro et al. [35] proposed the lightweight two-factor user authentication based on RSA cryptosystem for WSNs. However, the Watro et al. scheme [35] is vulnerable to replay, denial of service and key impersonation attacks [27, 28, 29, 30, 31]. Wong et al, [16] presented a lightweight user authentication scheme for WSNs, which only demands the computation of a hashing function. Later on, Srinivas et al. [22] show that the Wong et al. scheme [35] is vulnerable to stolen verifier and many logged-in users with the same login identity attack [27, 28, 29, 30, 31]. Tseng et al. [29] improved the version of Wong et al., which does not offer mutual authentication between the base-station and sensor-node [28]. To overcome the security weakness of mutual authentication, Lee et al. [30] presented a novel password based dynamic user authentication scheme, which also fails to satisfy mutual authentication between the base-station and the sensor-node [28].

Eschenauer et al. [32] presented a random based pre-distribution key mechanism to provide an initial trust between the sensor nodes. In random based key pre-distribution scheme, a key is randomly chosen from a key-pool and stored in the sensor node before it is deployed in the field. As a consequence, there are some certainties to have one common key for more than one sensor node. Chan et al. [33] improved this authentication scheme as two-key pre-distribution that has random pairwise-key and q-composite based key pre-distribution. Rasheed et al. [34] proposed a three-tier authentication scheme to provide a pairwise key establishment between the mobile-sink and the sensor-node. Nonetheless, the schemes, such as Chan et al. and Rasheed et al. have some serious security issues, namely user anonymity, intractability, privileged-insider and impersonation attack. Watro et al. [35] proposed an authentication scheme using Diffie-Hellman and RSA protocol as TinyPK scheme. But then, the TinyPK scheme is still susceptible to the masquerade attack [20]. To address this issue, Das et al. [20] introduced a two-factor user authentication scheme.

Chen et al. [37] shown that the Das et al. scheme is unsuccessful to provide the mutual authentication between the mobile-sinks and the sensor-nodes. To overcome the security weakness of Das et al. [23], Chen and Shih [36] proposed a novel authentication protocol for WSNs. He [24] extended the authentication scheme of Das et al. [23] to resist attacks such as privileged-insider and key impersonation. Yuan et al. [28] presented a biometric based user authentication scheme, which has a similar architecture of Das et al. scheme [23] to satisfy the security properties of AKA protocol. However, the Yuan et al. scheme is susceptible to denial of service (DoS) and node compromise attack.

Very few studies have focused on security issues for mobile-sink in WSNs [34, 38, 46, 47]. Let us assume that the adversary wants to impersonate as a legal mobile-sink to sense the most sensitive information from sensor-node or pretend as a legitimate sensor to upload incorrect or pseudo-kind of messages to the mobile-sink. Owing to mobility in the wireless enwvironment, most of the existing authentication schemes [20, 37, 39, 40, 41] are not well suited to authenticate the sensor-node and mobile-sink. As secure authentication scheme is believed to be essential to the mobile-sink in WSNs, this paper presents a novel seamless secure authentication scheme to improve the security efficiencies of the communication systems in terms of mutual authentication, session key agreement, user anonymity and intractability. The objectives of interaction and cooperation between the objects and the things are to send the data over wireless networks to signify the purpose of rapid development in the emerging technologies of IoT and cloud computing. In order to examine its common features and related discoveries, Stergiou et al. [48] presented a comparative study work, which focuses on the security issues of both the technologies. To provide the promising features, such as seamless interaction and interoperation, these technologies offer a smart home concept to associate the embedded computing technologies and network coverage. To solve the security and privacy preservation issues in the associated technologies, Tao et al. [49] presented a model of multilayer cloud. However, their architectural model fails to examine the mutual authentication and session key agreement between the communication entities.

Unlike the previous studies, S-SAKA framework tackles security issues, like data confidentiality, mutual authentication, session-key agreement, user anonymity, intractability and resilient to node-capture, key impersonation, password guessing and stolen smart-card attack for WSN configurations using the mobile-sink while providing seamless connectivity over WSNs to reduce the computation and communication overhead.

III. NETWORK MODEL AND ASSUMPTIONS

In this section, an architecture of a hierarchical WSN and the mathematical assumptions are discussed to signify the importance of the communication overhead and system security. The former is considered to mitigate the communication cost between the cluster head and the base station, whereby the network lifetime can be extended.

The latter is derived to provide a better security mechanism to protect the system under various potential attacks, such as privileged-insider, replay, stolen smart-card and node capture. The important notation used in proposed S-SAKA is illustrated in Table 10.1.

A. Network Model

The purpose of mobile-sink is to collect and upload the sensing data to the base-station. The principle use of mobile-sink is to mitigate the communication cost between cluster-head and base-station to enhance the network lifetime of the WSNs and reduce the communication overhead. The major disadvantage of the cloud-centric IoT is that it is usually based on a flat-topology structure that causes many problems such as scalability, increased traffic congestion among the nodes much closer to the sink (known as the broadcast storm), and an

Table 10.1 Important Notation Used in Proposed S-SAKA

NOTATION	DESCRIPTION
M_S	Mobile-sink
B_{St}	Base-station
U_{ser_j}	j^{th} user
PID_j	Identity of jth user
SK_j	Secret key of jth user
CH_i	i^{th} cluster head
CID_i	Unique identity of ith cluster head
$H_1 : \{0,1\}^*$	Map to point hashing functional operation
$H_2 : \{0,1\}^*$	Secure collision free one way cryptography hashing function
$\hat{e}$	Mapping function $G \times G \rightarrow G_T$
x	Secret random integer controlled by B_{St}
$E_{S_k}(.)$	Symmetric key encryption function
ΔTS	Expected delay transmission time
TS_S	Timestamp
$\|$	Concatenation operator
$\oplus$	Bitwise X-OR operator
$Data_j$	sensing data collected by CH
S, r, y, z	Random integers $\in Z_q^*$
LC_{DB_S}	Legal cluster-head database
P_{pub}	Public key
S_{k1}, S_{k2}	Secure session key
q	prime order integers

increase in overhead complexity. Therefore, clustering was introduced to subdivide the broadcast area into smaller cluster areas.

Many practical applications have had the model of hierarchical WSNs' for the purpose of power consumption; but then this paper is aimed to design a seamless secure authentication and key agreement (S-SAKA) under the architecture of hierarchical WSNs' to provide one-time user authentication. Figure 10.1 shows the architecture of the network model considered in this study.

Figure 10.1 shows that the sensor nodes are connected to cluster heads and cluster heads can communicate with each other as well as mobile-sinks. Dual way arrows on mobile sinks demonstrate the connections between sensor nodes and mobile sinks. For the cases where the communication is not within the range of the base station, the multi-hop data transmission can be employed; however, this can deteriorate the network lifetime and increase the communication overhead significantly. Hence, mobile-sinks can be employed which collects the sensed data from cluster head and upload it to the base station. The principle motivation behind the usage of mobile-sink in WSNs is to curtail the computation and communication overhead between the cluster head and the base station in order to enhance the WSNs lifetime.

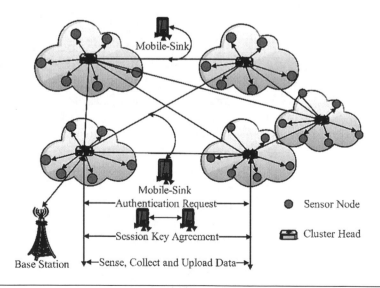

Figure 10.1 Architecture of hierarchical Cloud-based WSNs.

B. Mathematical Assumptions

In this subsection, the significance of Elliptic-Curve (EC) and Bilinear Pairing (BP) are introduced. In comparison with RSA, EC can provide better security level with minimum key length size [33].

Elliptic Curve: Assume that p is a prime number and f_p is a finite integer field with modulus p. Hence, an elliptic curve can be expressed as:

$$y^2 = x^3 + ax + b \,(mod\ p) \tag{10.1}$$

where $a, b \in f_p$ to satisfy the equation $4a^3 + 27b^2 \neq 0$. The scaling points $Q(x, y)$, which satisfies the above equation with ∞ is called as "Point at Infinity" to form an additive cyclic group as:

$$E[f_p] = \{(x, y): x, y \in f_p \text{ satisfy with } y^2 = x^3 + ax + b \,(mod\ p) \cup \infty\}.$$

In this aspect, the scalar multiplication of $Q(x, y)$ on EC can be computed with the repetitive addition of n, *i.e.*, $P = P + P + P ... + P(n\ times)$. The details of the assumptions can be found in [42, 43]. Table 10.1 shows the important notation used in proposed S-SAKA framework.

Bilinear Pairing: Assume that G is a cyclic (additive) group generated by a key-point P_K and G_T is a cyclic (multiplicative) group. The group parameters, such as G *and* G_T have same prime order q. Also, assume $\hat{e}: G \times G \rightarrow G_T$ be a computational bilinear mapping to satisfy the properties that are as follows:

Bilinearity: As sume, $X, Y \in G$ and $p, q \in Z_q^*$, $\hat{e}(pX, qY) = \hat{e}(X, Y)^{pq}$; also $Z_q^* = \{k \mid 1 \leq k \leq q - 1\}$.

Non-degenerate: Assume, $X \in G, \hat{e}(X, X) \neq e$, where e is the identity of the group element G_T.

Computability: Assume, $\hat{e}(X, Y)$ be an existing algorithm to compute the key-secrecy, for any $X, Y \in G$.

Mathematical Assumptions: To prove the importance of S-SAKA mechanism, some significant mathematical problems are derived from [44, 45] that are as follows:

- Discrete Logarithm (DL) Problem: Assume $(P, Q) \in G$ to find an integer $n \in Z_q^*$ such that $Q = nP$.

- Computational Diffie-Hellman (*CDH*) problem: Assume P, Px, Py, Pz for any random integer $x, y, z \in Z_q^*$ to determine xyP.
- Decisional Diffie-Hellman (*DDH*) problem: Assume P, Px, Py, Pz for any random integer $x, y, z \in Z_q^*$ to determine whether $zP = xyP$ or $z = xy \ (mod \ q)$.
- Bilinear Diffie-Hellman (BDH) problem: Assume P, Px, Py, Pz for any random integer $x, y, z \in Z_q^*$ to determine $\hat{e}(P, P)^{xyz}$.

IV. SEAMLESS SECURE AUTHENTICATION AND KEY AGREEMENT (S-SAKA) FRAMEWORK

In order to resolve the problem of security issue between mobile-sink and cluster-head, a framework of S-SAKA is proposed using bilinear-pairing. The S-SAKA framework is composed of seven phases: Initialization; System registration; Cluster-Head Registration; Mobile-Sink Registration; System Login; Authentication; Extraction of Sensing Data and secret key update. The mechanisms of S-SAKA are discussed in the following subsections.

A. Initialization Phase

In this phase, base station B_{St} performs an initialization to generate the prerequisite parameters keys to publish the system requirements. The procedural steps are as follows:

Step 1: Bilinear parameters are generated $\{q, P, G, G_T, \hat{e}\}$, where G is a cyclic additive group that generate prime order integers q by P and $\hat{e} : G \times G \to G_T$ is a bilinear group map.

Step 2: After the generation of bilinear parameters, B_{St} chooses a random integer $S \in Z_q^*$ as its corresponding master key to compute its public key p_{pub} using $p_{pub} = S.P$.

Step 3: After the generation of public key p_{pub}, B_{St} determines two secure collision resistance hashing operator H_1 and H_2, where $H_1 : \{0,1\}^*$ is called as map to point hashing functional operation and $H_2 : \{0,1\}^*$ is a one way secure hashing.

Step 4: After the determination of hashing function, B_{St} selects a symmetric encryption $E_{S_k}(.)$ as the system encryption function.

Step 5: Lastly, B_{St} publishes the system parameters $\{q, P, G, G_T, \hat{e}, P, p_{pub}, H_1, H_2, E_{S_k}(.)\}$ and keeps the secret parameters $\{S, x, y\}$ confidentially.

The procedural steps can be represented as an illustrative diagram as follows:

$$B_{St} \text{ performs an initialization}$$

$$G \text{ generates} \left\{ \begin{array}{l} q \; by \; P \\ \hat{e} : G \times G \rightarrow G_T \end{array} \right. \rightarrow \{q, P, G, G_T, \hat{e}\}$$

$$B_{St} \text{ chooses a random integer} : S \in Z_q^* \xrightarrow{Compute} p_{pub}$$

$$p_{pub} = S.P$$

$$B_{St} \text{ two secure collision resistance} : \left\{ \begin{array}{l} H_1 : \{0,1\}^* \\ H_2 : \{0,1\}^* \end{array} \right.$$

$$B_{St} \xrightarrow{\substack{symmetric \\ encryp}} E_{S_k}(.)$$

$$B_{St} \left\{ \begin{array}{l} sys \; para : \{q, P, G, G_T, \hat{e}, P, p_{pub}, H_1, H_2, E_{S_k}(.)\} \\ secret \; para : \{S, x, y\} \end{array} \right.$$

B. System Registration Phase

Prior to the deployment of the system, the available mobile sinks and cluster-head should be registered with the base-station. This phase is composed of two-parts, namely cluster-head and mobile-sink registrations. In the S-SAKA framework, cluster-head and mobile-sink have its own unique identity, as CID_i and PID_j respectively. Moreover, the base-station B_{St} has a database table DB_S which is initialized to maintain the non-compromised cluster-head.

1. Cluster-Head Registration   Before the deployment of cluster-head CH_i In WSNs', the base-station integrate a unique identity CID_i to compute $S.H_1(CID_i)$, and then, the base-station stores back the computed value of $S.H_1(CID_i)$ in to the memory of CH_i. Eventually, the base-station places CH_i at an appropriate position and insert a new identifier CID_i in the table of DB_S.

The cluster-head registration can also be represented as an illustrative diagram as follows:

$$B_{St} \text{ integrate } CID_I \xrightarrow{\ Compute\ } S.H_1(CID_i).$$

$$B_{St} \xrightarrow[\ S.H_1(CID_i)\ ]{\ Store\ } CH_i.$$

$$DB_S \xleftarrow[\ CID_i\ ]{} CH_i.$$

2. Mobile-Sink Registration   In this phase, the mobile-sink M_S which is authorized would try to store the random integer x into smart card. The procedural steps are as follow:

Step 1: The authorized mobile-sink can freely opt an identity PID_j, a secret-key SK_j and a random integer $x \in Z_q^*$. Afterwards, the mobile-sink determines $H_2(x \oplus SK_j)$ and sends the message-request $\{PID_j, H_2(x \oplus SK_j)\}$ to B_{St} via a secure communication channel.

Step 2: After receiving the message-request $\{PID_j, H_2(x \oplus SK_j)\}$, the B_{St} determines the following expressions: $Certify_j = S.H_1(PID_j \| H_2(x \oplus SK_j))$; $TS_j = H_2(PID_j \| y)$; $H_j = H_2(TS_j)$; $V_j = TS_j \oplus H_2(x \oplus SK_j))$; $A_j = H_2(PID_j \| x \| y)$. Then, B_{St} delivers a smart-card which is integrated of $\{Certify_j, V_j, H_j, A_j\}$ to M_S via a secure communication channel.

Step 3: After receiving the smart-card, the authorized mobile-sink stores back the random-integer x securely in the smart-card. Now, the smart-card is integrated of $\{Certify_j, V_j, H_j, A_j, x\}$.

The following illustrative diagram shows mobile-sink registration over base station.

$$M_S : \left\{ \begin{array}{c} PID_j \\ SK_j \\ x \in Z_q^* \end{array} \right.$$

$$\text{Commit} = H_2 \left(x \oplus SK_j \right)$$

$$M_S \xrightarrow{\{PID_j, \text{Commit}\}} B_{St}$$

$$\left. \begin{array}{c} Certify_j = S.H_1 \left(PID_j \parallel H_2 \left(x \oplus SK_j \right) \right) \\ TS_j = H_2 \left(PID_j \parallel y \right) \\ H_j = H_2 \left(TS_j \right) \\ V_j = TS_j \oplus H_2 \left(x \oplus SK_j \right) \\ A_j = H_2 \left(PID_j \parallel x \parallel y \right) \end{array} \right\} : B_{St}$$

$$M_S \xleftarrow[\{Certify_j, V_j, H_j, A_j\}]{Smart-card} B_{St}$$

$$M_S : \left\{ Certify_j, V_j, H_j, A_j, x \right\}$$

C. System Login Phase

When any user wishes to use M_S to sense data in an appropriate network, then the user should insert a valid smart-card into the terminal to enter the credentials, such as identity PID_j and secret-key SK_j. The execution flows are as follows:

Step 1: The smart-card S_C determines $TS_j^* = V_j \oplus H_2 \left(PID_j \parallel H_2 \left(x \oplus SK_j \right) \right)$ and $H_j^* = H_2 \left(TS_j^* \right)$ to verify whether H_j^* is equal to H_j or not. If the equality holds, then M_S executes the subsequent steps. Otherwise, S_C requests the user to provide the proper credentials, such as user identity and secret key.

Step 2: The smart-card determines a random-integer value rN_j to determine the following:

$$b_j = H_2\left(TS_j^* \| rN_j\right) \oplus H_2\left(x \oplus SK_j\right);$$

$$c_j = H_2\left(A_j \| H_2\left(x \oplus SK_j\right) \| rN_j\right),$$

then the mobile-sink sends the message-request $\{PID_j, b_j, c_j, rN_j\}$ to B_{St} via secure common channel.

Step 3: Upon receiving the message $\{PID_j, b_j, c_j, rN_j\}$ from smart-card, B_{St} performs the computation

$$A_j^* = H_2\left(PID_j \| x \| y\right);$$

$$D_j = b_j \oplus H_2\left(H_2\left(PID_j \| x\right) \| rN_j\right) = H_2\left(x \oplus SK_j\right);$$

$$c_j^* = H_2\left(A_j^* \| D_j \| rN_j\right).$$

After computation, B_{St} verifies whether $c_j^* = c_j$ or not. If the equality holds, then B_{St} determines M_S as a legal mobile-sink to compute

$$Q_j = H_2\left(\left(PID_j \| H_2\left(x \oplus SK_j\right)\right) \| rN_j\right),$$

$$M_K \ \hat{e}\left(S.H_1\left(PID_j \| H_2\left(x \oplus SK_j\right)\right), p_{pub}\right), \text{ and}$$

$$CLC_{DBs} = E_{MK}\left(LC_{DBs}\right).$$

Lastly, B_{St} sends the computation parameters $\{Q_j, CLC_{DBs}\}$ to M_S.

Step 4: Upon receiving the parameters $\{Q_j, CLC_{DBs}\}$ from B_{St}, M_S determines $Q_j^* = H_2\left(\left(PID_j \| H_2\left(x \oplus SK_j\right)\right) \| rN_j\right)$ to verify whether $Q_j^* = Q_j$ or not. If the equation holds, then M_S computes $M_K = \hat{e}\left(cert_j, p_{pub}\right)$ and $LC_{DBs} = D_{MK}\left(CLC_{DBs}\right)$, where D_{MK} is the decryption of $E_{MK}\left(.\right)$.

After the successful computation, M_S stores the updated list of legal cluster heads LC_{DBs} into its storage memory. Upon the successful of LC_{DBs}, M_S retains the complete database of legal cluster heads to run the subsequent step to ensure the authentication during message transmission. Moreover, to obtain the legal cluster head, the above steps are periodically executed.

The steps of system login phase can be presented as:

$$S_C \text{ determines}: \begin{cases} TS_j^* = V_j \oplus H_2\left(PID_j \parallel H_2\left(x \oplus SK_j\right)\right) \\ H_j^* = H_2\left(TS_j^*\right) \end{cases}$$

$$\text{if } H_j^* \neq H_j \text{ then}$$

$$S_C \xrightarrow[PID_j, \, SK_j]{requests} user$$

else

$$S_C \text{ } determines \text{ } rN_j : \begin{cases} b_j = H_2\left(TS_j^* \parallel rN_j\right) \oplus H_2\left(x \oplus SK_j\right) \\ c_j = H_2\left(A_j \parallel H_2\left(x \oplus SK_j\right) \parallel rN_j\right) \end{cases}$$

$$M_S \xrightarrow[\{PID_j, \, b_j, c_j, rN_j\}]{} B_{St}$$

$$\left. \begin{array}{c} A_j^* = H_2\left(PID_j \parallel x \parallel y\right) \\ D_j = b_j \oplus H_2\left(H_2\left(PID_j \parallel x\right) \parallel rN_j\right) = H_2\left(x \oplus SK_j\right) \\ c_j^* = H_2\left(A_j^* \parallel D_j \parallel rN_j\right) \end{array} \right\} : B_{St}$$

$$\text{if } c_j^* \neq c_j \text{ then}$$

$$M_S : \left\{ \begin{array}{c} Q_j = H_2\left(\left(PID_j \parallel H_2\left(x \oplus SK_j\right)\right) \parallel rN_j\right) \\ M_K \text{ } \hat{e}\left(S.H_1\left(PID_j \parallel H_2\left(x \oplus SK_j\right)\right), P_{pub}\right) \\ CLC_{DBs} = E_{M_K}\left(LC_{DBs}\right) \end{array} \right\} : B_{St}$$

$$M_S \xleftarrow[\{Q_j, CLC_{DBs}\}]{} B_{St}$$

$$M_S \text{ determines}: Q_j^* = H_2\left(\left(PID_j \parallel H_2\left(x \oplus SK_j\right)\right) \parallel rN_j\right)$$

$$\textit{if } Q_j^* = Q_j \text{ then}$$

$$M_K = \hat{e}\left(cert_j, p_{pub}\right)$$

$$LC_{DBs} = D_{M_K}\left(CLC_{DBs}\right)$$

D. System Authentication Phase

Once the system login phase is completed successfully, M_S can move into the WSN's coverage area to collect the sensing data. In order to authenticate its communication with CH_j, M_S has the procedural executions, which are as follows:

Step 1: While M_S actuates its current vicinity into CH_j, it transmits its connection-request to nearby CH_j for user authentication.

Step 2: After receiving the connection-request, CH_j sends its unique identity of CID_j to the requested mobile-sink.

Step 3: Upon receiving the CID_j, M_S verifies the legitimacy of CH_j using the un-compromised cluster database table LC_{DBs}. If CH_j is found as legal, then M_S generates the following: $H_t = H_2(TS_S)$; $m_1 = H_t.H_1(PID_j \parallel H_2(x \oplus SK_j)$; and $\Delta_j = \hat{e}\left(Certify_j.H_t.H_1(CID_j)\right)$. Then, the mobile-sink sends the message-request $\{m_1, \Delta_j, TS_S\}$ to CH_j. Eventually, the mobile-sink determines an initial session key $S_{K1} = \hat{e}\left(H_t.Certify_j, H_t.H_1(CID_j)\right)$ as a secret-session key.

Step 4: After the successful computation of $\{m_1, \Delta_j, H_t\}$, CH_j verifies whether $TS_S - TS_C \leq \Delta TS$, where TS_C is the current timestamp of CH_j message transmission and ΔTS is the expected transmission delay. If the delay interval is permissible, then CH_j determines

$$H_t = H_2(TS_S) \text{ and } \Delta_j^* = \hat{e}\left(Certify_j.H_t.H_1(CID_j)\right)$$

to check whether $\Delta_j^* = \Delta_j$ or not. If the equation holds, then CH_j identifies M_S to be legitimate. To create a common session key, CH_j computes a final session key

$$S_{k2} = \hat{e}\left(m_1, H_t.\mathfrak{r}.H_1\left(CID_j\right)\right)$$

and

$$\nabla = H_2\left(S_{k1} \| PID_j \| CID_j \| TS_S\right)$$

and then sends the message parameter $\{\nabla, TS_S\}$ to M_S.

To verify the identical in session key, M_S computes the equation that is as follows:

$$\begin{aligned}
S_{k2} &= \hat{e}\left(m_1, H_t.\mathfrak{r}.H_1\left(CID_j\right)\right) \\
&= \hat{e}\left(H_t.H_1\left(PID_j \| H_2(x \oplus SK_j)\right), H_t.\mathfrak{r}.H_1\left(CID_j\right)\right) \\
&= \hat{e}\left(H_t.\mathfrak{r}.H_1\left(PID_j \| H_2(x \oplus SK_j)\right), H_t.H_1\left(CID_j\right)\right) \\
&= \hat{e}\left(H_t.Certify_j, H_t.H_1\left(CID_j\right)\right) = S_{K1}
\end{aligned}$$

Step 5: Once the message parameter $\{\nabla, TS_S\}$ is received from CH_j, M_S determines $V_{erify} = H_2\left(S_{k1} \| PID_j \| CID_j \| TS_S\right)$ to check whether it holds with ∇ or not. If the verification is successful, M_S uses the session key S_{k1} to establish a session with CH_j.

The steps of system authentication phase can be presented as:

$$M_S \xrightarrow{\ connection-request\ } CH_j$$

$$\xleftarrow{\ CID_j\ }$$

$$M_S \ checks$$

$$CH_j \ in \ LC_{DB_S}$$

if CH_j is legal then

$$M_S : \begin{cases} H_t = H_2(TS_S) \\ m_1 = H_t.H_1(PID_j \| H_2(x \oplus SK_j) \\ \Delta_j = \hat{e}\left(Certify_j.H_t.H_1(CID_j)\right) \end{cases}$$

$$M_S \xrightarrow[\{m_1,\Delta_j,TS_S\}]{message-request} CH_j$$

$$S_{K1} = \hat{e}\left(H_t.Certify_j, H_t.H_1(CID_j)\right)$$

$$TS_S - TS_C \le \Delta TS\} : CH_j$$

$$\left.\begin{array}{c} H_t = H_2(TS_S) \\ \Delta_j^* = \hat{e}\left(Certify_j.H_t.H_1(CID_j)\right) \\ \# \end{array}\right\} : CH_j$$

if $\Delta_j^* = \Delta_j$ then

$$M_S \xleftarrow{identifies} CH_j$$

Computes

$$\left.\begin{array}{c} S_{k2} = \hat{e}\left(m_1, H_t.r.H_1(CID_j)\right) \\ \nabla = H_2\left(S_{k1} \| PID_j \| CID_j \| TS_S\right) \\ \# \end{array}\right\} : CH_j$$

$$M_S \xleftarrow{\{\nabla,TS_S\}} CH_j$$

$$S_{k2} = \hat{e}\left(m_1, H_t .r.H_1\left(CID_j\right)\right)$$

$$= \hat{e}\left(H_t .H_1\left(PID_j \parallel H_2(x \oplus SK_j)\right), H_t .r.H_1\left(CID_j\right)\right)$$

$$= \hat{e}\left(H_t .r.H_1\left(PID_j \parallel H_2(x \oplus SK_j)\right), H_t .H_1\left(CID_j\right)\right)$$

$$= \hat{e}\left(H_t .Certify_j, H_t .H_1\left(CID_j\right)\right) = S_{K1}$$

$$M_{S:} V_{erify} = H_2\left(S_{k1} \parallel PID_j \parallel CID_j \parallel TS_S\right)$$

E. Extraction of Sensing Data

While M_S is successfully established its communication with CH_j, it can read the sensing data from CH_j. The processing steps are as follows:

Step 1: Initially, CH_j computes the cipher text $CT_j = E_{S_{k2}}\left(Data_j\right)$ using the session key S_{k2}, where $Data_j$ is the internal storage data of CH_j. Lastly, CH_j sends the parameter CT_j to M_S.

Step 2: After receiving the parameter CT_j from CH, M_S determines $CData_j = E_{MK}\left(D_{S_{k1}}\left(CT_j\right)\right)$ to stores its corresponding values in internal storage device or sends the value to B_{St} directly, where $M_K = \hat{e}\left(cert_j, p_{pub}\right)$. The aforesaid equation may be inferred as: $CData_j = E_{MK}\left(D_{S_{k1}}\left(CT_j\right)\right) = CData_j = E_{MK}\left(D_{S_{k1}}\left(E_{S_{k2}}\left(Data_j\right)\right)\right) = E_{MK}\left(Data_j\right)$. Hence, the verification proves that the storage data of M_S is encrypted by M_K.

Step 3: After successful verification, M_S sends the transmission message $\left\{PID_j, CData_j, b_j, c_j, rN_j\right\}$ to B_{St}, where b_j, c_j and rN_j are computed in system login phase.

Step 4: Upon receiving the transmission message from M_S, B_{St} tries to extract PID_j to compute the equations, which are as follows: $A_j^* = H_2\left(PID_j \parallel x \parallel y\right); \quad D_j = H_2\left(x \oplus SK_j\right)$ and $c_j^* = H_2\left(A_j^* \parallel D_j \parallel rN_j\right)$.

After the successful computation, B_{St} verifies whether $c_j^* = c_j$ or not. If the equation holds, then B_{St} determines M_S to be a legitimate. Then, B_{St} determines $M_K = \hat{e}\left(S.H_1\left(PID_j \parallel H_2\left(x \oplus SK_j\right)\right), p_{pub}\right)$ to decrypt the storage- data $D_{MK}\left(Data_j\right) = D_{MK}\left(E_{MK}\left(Data_j\right)\right) = Data_j$. After decryption of storage data, B_{St} is allowed to extract the sensing data $Data_j$ from M_S.

The following diagram shows extraction of sensing data.

$$CT_j = E_{S_{k2}}\left(Data_j\right)\} : CH_j$$

$$M_S \xleftarrow{\quad CT_j \quad} CH_j$$

M_S determines

$$CData_j = E_{MK}\left(D_{S_{k1}}\left(CT_j\right)\right)$$

$$M_S \xrightarrow{\overset{sends}{M_K = \hat{e}\left(cert_j, p_{pub}\right)}} B_{St}$$

$$CData_j = E_{MK}\left(D_{S_{k1}}\left(CT_j\right)\right) = CData_j$$
$$= E_{MK}\left(D_{S_{k1}}\left(E_{S_{k2}}\left(Data_j\right)\right)\right) = E_{MK}\left(Data_j\right)$$

$$M_S \xrightarrow{\left\{PID_j, CData_j, b_j, c_j, rN_j\right\}} B_{St}$$

$$\left.\begin{array}{c} A_j^* = H_2\left(PID_j \,\|\, x \,\|\, y\right) \\ D_j = H_2\left(x \oplus SK_j\right) \\ c_j^* = H_2\left(A_j^* \,\|\, D_j \,\|\, rN_j\right) \end{array}\right\} : B_{St}$$

if $c_j^* = c_j$ *then* B_{St} determines M_S

$$M_K = \hat{e}\left(S.H_1\left(PID_j \,\|\, H_2\left(x \oplus SK_j\right)\right), p_{pub}\right)$$

$$D_{MK}\left(Data_j\right) = D_{MK}\left(E_{MK}\left(Data_j\right)\right) = Data_j$$

F. Secret Key Update Phase

In this phase, U_{ser} can modify his / her secret key when he / she wants to change. The procedural steps of key update phase are as follows:

Step 1: U_{ser} tries to enter his / her smart card into the user terminal to verify the credentials, such as PID_j and SK_j.

Step 2: After the successful entries, the smart card computes $TS_j^* = Ver_j \oplus H_2\left(PID_j \parallel H_2\left(x \oplus SK_j\right)\right)$ and $H_j^* = H_2\left(TS_j^*\right)$ to verify whether $H_j^* = H_j$ or not. If the equation holds, then U_{ser} is permitted to change his /her secret key SK_j^{new} and x^{new}. Otherwise, the smart card disapproves the request of U_{ser}. Lastly, M_S sends the transmission message $\left\{PID_j \parallel H_2\left(x \oplus SK_j\right), Ver_j, H_2\left(x^{new} \oplus SK_j^{new}\right)\right\}$ to B_{St} through secure communication channel.

Step 3: After receiving the transmission message $\left\{PID_j \parallel H_2\left(x \oplus SK_j\right), Ver_j, H_2\left(x^{new} \oplus SK_j^{new}\right)\right\}$ from M_S, B_{St} determines $Ver_j^* = H_2\left(PID_j \parallel y\right) \oplus H_2\left(PID_j \parallel H_2\left(x \oplus SK_j\right)\right)$ to verify whether $Ver_j^* = Ver_j$ or not. If the equation holds, then B_{St} performs a computation of $cert_j^{new} = r.H_1\left(PID_j\right) \parallel H_2\left(x^{new} \oplus SK_j^{new}\right)$ and $Ver_j^{new} = H_2\left(PID_j \parallel y\right) \oplus H_2\left(PID_j \parallel H_2\left(x^{new} \oplus SK_j^{new}\right)\right)$. Then, B_{St} sends the computation message $\left\{cert_j^{new}, Ver_j^{new}\right\}$ to M_S through a secure communication channel.

Step 4: After receiving the message $\left\{cert_j^{new}, Ver_j^{new}\right\}$ from B_{St}, the smart card modifies the parameters, such as $cert_j, verf_j$ and x into $cert_j^{new}, Ver_j^{new}$ and x^{new} in the given order.

The following illustrative diagram shows secret key update phase.

$$U_{ser} \text{ verify the credentials } PID_j \text{ and } SK_j$$

$$\left.\begin{array}{c} TS_j^* = Ver_j \oplus H_2\left(PID_j \parallel H_2\left(x \oplus SK_j\right)\right) \\ H_j^* = H_2\left(TS_j^*\right) \end{array}\right\} : \text{smart card}$$

$$if \ H_j^* = H_j \ then$$

$$U_{ser} \text{ is permitted to change his/her secret key } SK_j^{new} \text{ and } x^{new}$$

$$M_S \xrightarrow{\left\{ PID_j \| H_2\left(x \oplus SK_j\right), Ver_j, H_2\left(x^{new} \oplus SK_j^{new}\right)\right\}} B_{St}$$

$$Ver_j^* = H_2\left(PID_j \| y\right) \oplus H_2\left(PID_j \| H_2\left(x \oplus SK_j\right)\right) \Big\} : B_{St}$$

$$\text{if } Ver_j^* = Ver_j \text{ then}$$

$$\left. \begin{array}{l} cert_j^{new} = r.H_1\left(PID_j\right) \| H_2\left(x^{new} \oplus SK_j^{new}\right) \\ Ver_j^{new} = H_2\left(PID_j \| y\right) \oplus H_2\left(PID_j \| H_2\left(x^{new} \oplus SK_j^{new}\right)\right) \end{array} \right\} : B_{St}$$

$$M_S \xrightarrow{\left\{ cert_j^{new}, Ver_j^{new}\right\}} B_{St}$$

$$M_S \text{ modifies} : \left\{ \begin{array}{c} cert_j, verf_j = cert_j^{new}, Ver_j^{new} \\ x = x^{new} \end{array} \right.$$

V. SECURITY ANALYSIS

This section is composed of stringent formal and informal security analysis of S-SAKA. The analysis result shows that the proposed S-SAKA framework not only offers security properties of authentication protocols for mutual authentication, session-key agreement and data confidentiality, but also prevents the various potential attacks, such as node-capture, stolen smart-card, key impersonation and privileged-insider.

A. Formal Security Analysis

S-SAKA framework offers secret session-key agreement between a legal cluster head CH_i, base station B_S, smart card S_C and a mobile-sink MS_i and it is proven using BAN logic [19]. Assume that X *and* Y be the principles, P *and* Q be the statement / formula and s_k be the secret key. The important notation used in the BAN logic is given in Table 10.2.

The BAN logic postulates are as follows:

Table 10.2 Important Notation Used in Ban Logic

NOTATION	DESCRIPTION
$X \models P$:	X relies on a statement of P
$\neq P$:	P be sure as fresh
$X \Longmapsto P$:	X takes the jurisdiction over P
$X \triangleleft P$:	X realizes P
$X \mid \sim P$:	X formerly believed as P
(P,Q):	PoQ is an individual part of (P,Q)
$\{P\}_{s_k}$:	P is encrypted using secret ket s_k
$\langle P \rangle_{s_k}^{Q}$:	P is mutually shared with Q
$X \overset{s_k}{\longleftrightarrow} Y$:	X and Y uses a secret-key s_k to establish a communication. Besides, s_k is totally secure; and thus can not be discovered by any principal excluding X and Y.

Rule 1 – Meaning of Messages:

$$\frac{X \models X \overset{s_k}{\longleftrightarrow} Y, X \triangleleft \{P\}s_k}{X \models Y \mid \sim P} \quad \text{and} \quad \frac{X \models X \overset{Q}{\longleftrightarrow} Y, X \triangleleft \{P\}_Q}{X \models Y \mid \sim Q} : \text{ If } X$$

trusts that s_k is shared among X and Y and perceives P encrypted with s_k, then X trusts the Y as a legal client.

Rule 2 – Verification of Nonce:

$$\frac{X \models \neq P, X \models Y \mid \sim P}{X \models Y \models P} \quad \text{and} \quad \frac{X \models \neq Q, X \models Y \mid \sim Q}{X \models Y \models Q} : \text{ If } X \text{ trusts that } X$$

has just been communicated and thus Y ony perceives P, then X trusts that Y be certain of P.

Rule 3 – Belief:

$$\frac{X \models P \quad X \models Q}{X \models (P, \quad Q)} : \text{ If } X \text{ trusts } P \text{ and } Q, \text{ then } X \text{ beliefs in } P \text{ and } Q.$$

Rule 4 – Rule of Fresh-Concatenation:

$$\frac{X \models \neq P}{X \models \neq (P, \quad Q)} : \text{ If } X \text{ trusts the freshness in key generation of } P,$$

then Y be certain of freshness in (P,Q).

Rule 5 – Rule of Jurisdiction:

$$\frac{X \models Y \Longrightarrow P \quad X \models Y \models P}{X \models (P, \quad Q)} : \text{ If } X \text{ trusts that } Y \text{ has influence over}$$

P and X believes Y in the accuracy of P, then X trusts in P.

To satisfy the security properties of AKA protocol, the proposed S-SAKA framework must be able to meet all the test goals, given in below.

$$Goal_1 : M_{S_i} |\equiv B_S| \equiv CH_i \overset{sk}{\leftrightarrow} S_C$$

$$Goal_2 : M_{S_i} |\equiv CH_i \overset{sk}{\leftrightarrow} S_C$$

$$Goal_3 : S_C |\equiv M_{S_i}| \equiv CH_i \overset{sk}{\leftrightarrow} S_C$$

$$Goal_4 : S_C |\equiv CH_i \overset{sk}{\leftrightarrow} S_C$$

The structural flow of BAN logic is as follows:

1. Messages in Generic Form:

$$M_1 : M_{S_i} \rightarrow B_S : \left\langle H_2\left(x \oplus SK_j\right)\right\rangle , \left\langle PID_j, H_2(x \oplus SK_j\right\rangle_{x \in Z_q^*}$$

$$M_2 : B_S \rightarrow S_C :$$
$$\left\langle Certify_j = S.H_1\left(PID_j \| H_2\left(x \oplus SK_j\right)\right);\right.$$
$$TS_j = H_2\left(PID_j \| y\right); H_j = H_2\left(TS_j\right);$$
$$V_j = TS_j \oplus H_2\left(x \oplus SK_j\right));$$
$$\left.A_j = H_2\left(PID_j \| x \| y\right)\right\rangle_{x \in Z_q^*}$$

$$M_3 : M_{S_i} \rightarrow S_C : \left\langle Certify_j, V_j, H_j, A_j, x\right\rangle_{x \in Z_q^*}$$

2. Transmission of Messages in Idealized Form:

$$T_{M1} : M_{S_i} \rightarrow B_S : \left\langle PID_j, H_2(x \oplus SK_j\right\rangle_{MS_i \overset{PID_j}{\rightarrow} B_S}$$

$$T_{M2} : B_S \rightarrow S_C : \left\langle Certify_j, V_j, H_j, A_j\right\rangle_{MS_i \overset{PID_j}{\rightarrow} B_S}$$

$$T_{M3} : M_{S_i} \rightarrow S_C : \langle Certify_j, V_j, H_j, A_j, x \rangle_{MS_i \xrightarrow{PID_j} B_S}$$

3. Messages in Hypotheses Form:

$$H_{M1} : M_{S_i} \mid\equiv\neq (CID_i), \; CH_i \mid\equiv\neq (TS_1, TS_2)$$

$$H_{M2} : B_S \mid\equiv\neq (PID_i), \; B_S \mid\equiv\neq (TS_3, TS_4)$$

$$H_{M3} : M_{S_i} \mid\equiv B_S \mid\equiv CH_i \overset{sk}{\leftrightarrow} S_C$$

$$H_{M4} : M_{S_i} \mid\equiv CH_i \overset{sk}{\leftrightarrow} S_C$$

$$H_{M5} : S_C \mid\equiv MS_i \mid\equiv CH_i \overset{sk}{\leftrightarrow} S_C$$

$$H_{M6} : S_C \mid\equiv CH_i \overset{sk}{\leftrightarrow} S_C$$

The idealized form of S-SAKA framework is examined on the base of BAN logic postulates and goal settings. The key proofs of S-SAKA are as follows:

- From the transmission message T_{M1}, the S-SAKA has $P_1 : B_S \lhd \langle PID_j, H_2(x \oplus SK_j) \rangle_{MS_i \xrightarrow{PID_j} B_S}$.

- From H_{M2}, P_1 and $Rule(1)$, the S-SAKA acquires $P_2 : B_S \mid\equiv M_{S_i} \mid\sim \langle PID_j, H_2(x \oplus SK_j) \rangle$.

- From the transmission message T_{M2}, the S-SAKA has $P_3 : S_C \lhd \langle Certify_j, V_j, H_j, A_j \rangle_{MS_i \xrightarrow{PID_j} B_S}$.

- From H_{M5}, P_3 and $Rule(1)$, the S-SAKA acquires $P_4 : CH_i \mid\equiv S_C \mid\sim \langle Certify_j, V_j, H_j, A_j \rangle$.

- From H_{M1}, P_4, $Rule(2)$ and $Rule(4)$, the S-SAKA obtains $P_5 : M_{S_i} \mid\equiv B_S \mid\equiv CH_i \overset{sk}{\leftrightarrow} S_C . \langle \mathbf{Goal}_1 \rangle$

- Try, from H_{M5}, P_5 and $Rule(1)$, the S-SAKA gets $P_6 : M_{S_i} \mid\equiv CH_i \overset{sk}{\leftrightarrow} S_C . \langle \mathbf{Goal}_2 \rangle$

- From the transmission message T_{M3}, the S-SAKA has
  $P_7 : B_S \vartriangleleft \langle Certify_j, V_j, H_j, A_j, x \rangle_{MS_i \overset{PID_j}{\to} B_S}$.

- From H_{M1}, P_7 and $Rule(1)$, the S-SAKA acquires
  $P_8 : B_S \models MS_i | \sim \langle Certify_j, V_j, H_j, A_j, x \rangle$.

- From H_{M5}, P_8, $Rule(2)$ and $Rule(4)$, the S-SAKA obtains
  $P_9 : S_C \models M_{S_i} | \equiv CH_i \overset{s_k}{\leftrightarrow} S_C . \langle \boldsymbol{Goal}_3 \rangle$

- From H_{M4}, P_9 and $Rule(3)$, the S-SAKA achieves
  $P_{10} : S_C \models CH_i \overset{s_k}{\leftrightarrow} S_C . \langle \boldsymbol{Goal}_4 \rangle$

- Again, from H_{M6}, P_{10} and $Rule(5)$, the S-SAKA produces
  $P_{11} : M_{S_i} \models CH_i \overset{s_k}{\leftrightarrow} S_C . \langle \boldsymbol{Goal}_2 \rangle$

Provided the goals $\boldsymbol{Goal}_1 - \boldsymbol{Goal}_4$, the S-SAKA protocol asserts that it uses a shared secret-key s_k to establish a communication; and hence the proposed S-SAKA framework is proficient to achieve the proper mutual authentication, session-key agreement and confidentiality.

B. Informal Security Analysis

In this subsection, the informal security analysis of S-SAKA protocol is performed in which the adversary has some unique capabilities that are as follows:

1. The adversary is able to control over the communication channel especially with mobile-sink, cluster-head and base-station to do message intercept, insert, delete or modify any exchange of information.

2. The adversary may incur either user identity and secret key or the storage information of smart card but he / she cannot obtain both. For an instance, if the adversary obtains the user identity and secret key, he / she can't have any chance to obtain the storage information of smart card.

1. Proper Mutual Authentication and Session-Key Agreement In the authentication phase, the cluster-head CH_j and mobile-sink authenticate each other by the verification of $\Delta_j = \hat{e}\left(Certify_j . H_t . H_1\left(CID_j\right)\right)$

to validate the secret-session key $S_{K1} = \hat{e}\left(H_t.Certify_j, H_t.H_1\left(CID_j\right)\right)$. Using $S.H_1\left(CID_j\right)$, the cluster-head performs the computation, which is as follows:

$$\hat{e}\left(m_1, H_t.r.H_1\left(CID_j\right)\right)$$
$$= \hat{e}\left(H_t.H_1\left(PID_j \parallel H_2(x \oplus SK_j)\right), S.H_1\left(CID_j\right)\right)$$
$$= \hat{e}\left(S.H_1\left(PID_j \parallel H_2(x \oplus SK_j)\right), H_t.H_1\left(CID_j\right)\right)$$
$$= \hat{e}\left(Certify_j, H_t.H_1\left(CID_j\right)\right) = \Delta_j$$

On the other hand, the mobile-sink authenticates the cluster-head using $Ver_j = \nabla$ to render the transmission message $\{\nabla, TS_S\}$. As the certificate authorization is given only for the authorized mobile-sink, the other cannot infer / forge to generate a valid authentic key value Δ_j. To establish a secure communication, the mobile-sink and the cluster-head shares s session key $S_{K1} = \hat{e}\left(H_t.Certify_j, H_t.H_1\left(CID_j\right)\right)$.Hence, the S-SAKA framework provides proper mutual authentication and session-key agreement.

2. *Data-Confidentiality* In S-SAKA framework, to collect the sensing data, the mobile-sink should try to achieve the proper mutual authentication with cluster-head using shared session key. After the establishment of session key, the mobile-sink can acquire the sensing information through the knowledge of cluster-head. As the shared session key is kept secretly between the mobile-sink and cluster-head, the adversary cannot deduce the $Data_j$ in plaintext. Thus, the S-SAKA framework claims that it can provide a secure communication between the mobile-sink and the cluster-head.

On the one hand, the secret-session key $S_{K1} = \hat{e}\left(H_t.Certify_j, H_t.H_1\left(CID_j\right)\right)$ is interfaced between the cluster-head CH and the mobile-sink. On the other hand, the cluster-head determines $CData_j = E_{MK}\left(D_{S_{k1}}\left(CT_j\right)\right) = CData_j$ to save and send it to the base-station. Even if the adversary acquires the information of mobile-sink, he / she cannot infer $Data_j$ as it could not obtain the key value of M_k. As the adversary, cannot tamper the sensing-data without knowledge of S_K, the S-SAKA framework provides data-confidentiality for users.

3. Resilient to Node-Capture Attack The resistance of node-capture attack can be measured effectively with the elimination of network communication, which are compromised by 'N' captured nodes directly [36]. Owing to inattentive property of WSNs', an adversary may capture the information of sensor-node or cluster-head. For your kind note, cluster-head has authentic identity CID_j and secret key value $S.H_1(CID_j)$ in initialization phase. Consequently, the adversary may have a chance to compromise the nodes, which are yet to communicate with mobile-sink and cluster-head. But then, the nodes, which are not compromised are still secure to establish the communication between mobile-sink and cluster-head. Subsequently, the S-SAKA framework claims that the adversary cannot provide any security disruption for uncompromised cluster-head and mobile-sink.

According to the reference [42, 43], the mobile-sink owner can deduce the recent updated cluster-head from LC_{DB_S} database, as soon as he / she has successfully logged into the base-station. Similarly, the mobile-sink can identify the compromised cluster-head timely to reject the compromised cluster-head. The un-compromised database table DB_S is associated with the mobile-sink securely; and thus the adversary cannot affect / damage the secure communication between the cluster-head and mobile-sink. Hence, the S-SAKA framework is resilient to node-capture attack.

4. Resilient to Stolen Smart-Card Attack Assume that adversary obtains the smart-card of the user MS_j; and thus he / she acquires the details of $Certify_j = S.H_1(PID_j \| H_2(x \oplus SK_j))$, $TS_j = H_2(PID_j \| y)$, $H_j = H_2(TS_j)$, $V_j = TS_j \oplus H_2(x \oplus SK_j)$, $A_j = H_2(PID_j \| x \| y)$. But then, the adversary can not deduce the users' unique identity PID_j and secret-key SK_j from $Certify_j, V_j, H_j$ and A_j owing to one-way property of the hash function $H_1(.)$ and $H_2(.)$. Therefore, the adversary cannot compute a precise $m_1 = H_t.H_1(PID_j \| H_2(x \oplus SK_j))$ to form a valid request message $\{m_1, \Delta_j, TS_S\}$. Therefore, the S-SAKA claims that it is resilient to stolen smart-card attack.

5. *Resilient to Replay Attack* Using replay attack, the adversary uses a falsified authentication process to acquire the system access. In order to deduce such false assumption, the S-SAKA uses timestamp TS_S. Assume an adversary wishes to launch a replay attack to infer the sensing data from cluster head CH. To extract the sensed data, the adversary needs to send an authentic message to CH. If the message $\{m_1^*, \Delta_j^*, TS_s^*\}$ is found to be expired or already used by another mobile-sink M_S, CH determines to be a susceptible behavior. Even though, the adversary changes the timestamp TS_s^*, he / she cannot find a proper Δ_j^* without the key parameter of B_{st} value S.

On the other hand, an adversary may wish to launch a replay attack to intercept with authentic mobile-sink M_S. To establish the communication, the adversary need to generate an authentic message $\{\nabla, TS_S\}$. As the key parameter of B_{st} value S is always kept secret between M_S and B_{st}, the adversary cannot determine a valid secret session key $S_{k2} = \hat{e}\left(m_1, H_t .r. H_1\left(CID_j\right)\right)$. Hence, the S-SAKA claims that the adversary cannot launch a replay attack without the proper computation of $\nabla = H_2\left(S_{k1} \| PID_j \| CID_j \| TS_S\right)$. This proves that the S-SAKA framework is resilient to replay attack.

6. *Resilient to Key Impersonation Attack* By using this attack, an adversary provides a forged information $\{m_1^*, \Delta_j^*, TS_s^*\}$ to impersonate as a legitimate mobile-sink M_S as to overhear the sensing information. However, the adversary cannot infer / forge Δ_j^* without the determination of $Certify_j$. According to DL problem, it is very much difficult to derive the secret key parameter S using P and p_{pub}. As a result, the adversary cannot determine $S.H_1\left(PID_j \| H_2\left(x \oplus SK_j\right)\right) = Certify_j$ and Δ_j^*. The above analysis proves that S-SAKA framework is resilient to key impersonation attack.

7. *Resilient to Privileged-Insider Attack* In the system registration phase of S-SAKA framework, the mobile-sink owner U_{ser} does not share his/her secret key SK_j in plaintext form. But, he / she shares its information as $H_2\left(x \oplus SK_j\right)$ to B_{st}. As $H_2(.)$ is a one-way point secure hashing function, it is computationally not possible to obtain SK_j. Moreover, the administrator or privileged B_{st} cannot determine a valid SK_j of U_{ser} and thus he / she cannot impersonate as a legal user U_{ser} to communicate with CH. Hence, the

S-SAKA framework claims to be secure against the privileged-insider attack.

8. User Anonymity and Intractability   In the system authentication phase, the S-SAKA framework uses mobile-sink M_S to send the transmission message $\{m_1, \Delta_j, TS_S\}$ to cluster-head *CH* in turn to obtain a proper user authentication. As each message transmission has unique time stamp TS_S that traverses between M_S and *CH*, there will be no correlation of two authentic messages, namely $\{m_1, \Delta_j, TS_S\}$ and $\{m_1^*, \Delta_j^*, TS_s^*\}$. Moreover, as the message transmission has one way point to map hashing function, it is much difficult to retrieve PID_j from m_1. Hence, the S-SAKA framework claims that the adversary cannot identify any authentic mobile-sink or communication link launched by the same mobile-sink.

9. Resilient to Offline Password-Guessing Attack   This attack is categorized into two cases that are as follows:

Case 1: Assume an insider wishes to know the information of legitimate user, such as user identity PID_j and secret-key SK_j during system registration. The registration request of insider $\{PID_j, SK_j)\}$ is sent securely to the base-station. Besides, the insider has a smart-card, which are stolen from U_{ser}. Even though he has the device access and user information, he / she could not derive a proper secret session key without the knowledge of secret key value x.

Case 2: Assume an outsider has stolen the smart-card of U_{ser}. As a consequence, he / she can extract all the confidential information of smart-card, such as $\{PID_j, b_j, c_j, rN_j\}$, where

$$b_j = H_2\left(TS_j^* \| rN_j\right) \oplus H_2\left(x \oplus SK_j\right),$$

$$c_j = H_2\left(A_j \| H_2\left(x \oplus SK_j\right) \| rN_j\right).$$

To derive a secret key SK_j, the outsider needs to know secret key value x, which is a bilinear parameter corresponding to Z_p^*. As it is controlled and changed its value periodically by the base-station, the outsider cannot guess the proper secret key SK_j to gain the U_{ser} access.

The above analysis proves that the proposed S-SAKA framework can be resilient to offline password-guessing attack.

10. Resilient to DoS Attack Without proper user identity PID_j and secret key SK_j, none of the user can successfully log in to the systems. Even if they have stolen the smart card of legitimate user, they can infer the information like $\{PID_j, b_j, c_j, rN_j\}$, where $b_j = H_2(TS_j^* \| rN_j) \oplus H_2(x \oplus SK_j)$, $c_j = H_2(A_j \| H_2(x \oplus SK_j) \| rN_j)$. After that, the smart-card verifies whether $H_j^* = H_2(TS_j^*)$ is valid or not, where $H_j = H_2(TS_j)$, $TS_j = H_2(PID_j \| y)$ and $TS_j^* = V_j \oplus H_2(PID_j \| H_2(x \oplus SK_j))$. As the timestamp TS_j and secret value x periodically changes, they cannot derive a proper secret key SK_j to gain the user access. Hence, the proposed S-SAKA framework is resilient to denial of service attack.

11. Resilient to Many Logged-In Users with the Same Login Identity Attack In the proposed S-SAKA framework, the user must provide valid credentials $\{PID_j, SK_j\}$ to obtain the access of cluster-head through the knowledge of base-station, which verifies the secret value x to authorize the service access. As the secret value x is unique to U_{ser} and controlled by base-station, the user redundancy cannot be determined using following expressions: $Certify_j = S.H_1(PID_j \| H_2(x \oplus SK_j))$; $TS_j = H_2(PID_j \| y)$; $H_j = H_2(TS_j)$; $V_j = TS_j \oplus H_2(x \oplus SK_j)$; $A_j = H_2(PID_j \| x \| y)$ as it is already in use. Hence, the proposed S-SAKA framework claim that it is resilient to many to many logged-in users with the same login identity attack.

VI. PERFORMANCE EVALUATION

In this section, the proposed S-SAKA framework is evaluated and compared with its related authentication schemes. To better understand the evaluation criteria of communication cost, some notation is defined as follows:

T_{SH} is defined as the execution time of one-way secure hashing function $H_2(.)$. T_{MH} is defined as the execution time of one-way point to map hashing function $H_1(.)$.

T_P is defined as the computation time of bilinear pairing function.

T_A is defined as the execution time of one-point additional operational function.

T_{ED} is defined as the execution time of encryption and decryption algorithmic function.

T_M is defined as the execution time of elliptic-curve scalar multiplication function.

In WSNs, energy efficiency is a major constraint and thus a lightweight user authentication protocols are preferred to mitigate the computational cost of the systems.

In order to reduce the amount of computations required, the proposed S-SAKA protocol uses cost inexpensive operations like hashing function and less cost expensive operation, such as bilinear pairing, encryption/decryption and scalar multiplication operation. To evaluate the cryptographic operations employed, an extensive verification is performed using MIRACLE C/C++ library with the system features of 32-bit Windows 7 Operating Systems and Microsoft Visual C++. To examine realistically, the execution time of symmetric key encryption/decryption $(AES-128)$, elliptic-curve point scalar multiplication over finite-field f_p and $SHA-1$ hashing function are set as $T_P \approx 0.0001\ ms$, $T_{ED} \approx 0.1303\ ms$, $T_M \approx 7.3529\ ms$ and $T_{SH} \approx T_{MH} \approx 0.0004\ ms$ as referred in [22]. Table 10.3 demonstrates the communication efficiencies of the proposed S-SAKA and its related existing authentication schemes [17, 18, 19, 20, 21, 22] during system login and authentication phase. Results show that, the computation cost of the bilinear pairing and scalar multiplication of proposed S-SAKA is comparatively short.

The examination results prove that the proposed S-SAKA has less communication overhead as it does not invoke the base-station to authenticate the mobile-sink and sensor node except during the secure communication establishment to provide seamless connectivity. Thus, it can be well suitable for WSN's environment in relation with the existing authentication schemes [17, 18, 19, 20, 21, 22].

Table 10.3 Comparison of Communication Efficiencies During System Login and Authentication Phase

AUTHENTICATION SCHEMES	MOBILE-SINK	CLUSTER-HEAD	BASE-STATION	TOTAL COST	EXECUTION TIME (MS)
Deebak [17]	$9T_{MH}$	$5T_{MH}$	$12T_{MH}$	$26T_{MH}$	0.0104
Turkanovic et al. [18]	$5T_{MH}$	$7T_{MH}$	$7T_{MH}$	$19T_{MH}$	0.0076
Farash et al. [19]	$11T_{MH}$	$7T_{MH}$	$14T_{MH}$	$32T_{MH}$	0.0128
Das et al. [20]	$9T_{MH}+1T_{ED}$	$3T_{MH}+1T_{ED}$	$5T_{MH}+2T_{ED}$	$17T_{MH}+4T_{ED}$	1.2480
Amin [21]	$7T_{MH}$	$5T_{MH}$	$8T_{MH}$	$20T_{MH}$	0.0080
Srinivas et al. [22]	$10T_{MH}$	$6T_{MH}$	$13T_{MH}$	$29T_{MH}$	0.0116
Proposed S-SAKA	$10T_{MH}+2T_P$	$5T_{MH}+2T_P$	$14T_{MH}+1T_P+3T_M$	$29T_{MH}+5T_P+3T_M$	0.0064 [Mobile-Sink and Cluster-Head Only]

Table 10.3 compares the communication overhead involved in system login and authentication phases for proposed and other existing schemes [17, 18, 19, 20, 21, 22]. While using SIIA-1 hashing, the one-way hash function is assumed to be $160 - bits \left[20\ bytes\right]$. In addition, for each random nonce, the identity of sensor node is set to be $152 - bits\left[19\ bytes\right]$. In proposed S-SAKA, during system login phase, the login message transmission request $T_{Msg1} = \{PID_j,\ b_j, c_j, rN_j\}$ and $T_{Msg2} = \{Q_j, CLC_{DBs}\}$ involves 78 *bytes* and 39 *bytes* respectively. In the course of system authentication and key agreement phase, the message transmission request $T_{Msg3} = \{m_1, \Delta_j, TS_S\}$ and $T_{Msg4} = \{PID_j, CData_j, b_j, c_j, rN_j\}$ encompasses 58 *bytes* and 98 *bytes*. As a result, during system login and authentication phase, the communication overhead is cumulated as follows: $\left[78 + 39 + 58 + 98\right] = 273 bytes$. On the other hand, the communication overheads involved in system login and authentication phases for Deebak [17], Turkanovic et al. [18], Farash et al. [19], Das et al. [20], Amin [21] and Srinivas et al. [22] are calculated as 315 *bytes*, 489 *bytes*, 434 *bytes*, 391 *bytes*, 373 *bytes* and 353 *bytes* respectively. It is observed that the proposed S-SAKA scheme provides less communication overhead in comparison with other existing authentication schemes [17, 18, 19, 20, 21, 22].

Table 10.4 Assertion Threshold Test Values

PARAMETER	VALUE
g_{t1}	0.00
g_{t2}	0.33
g_{t3}	0.66

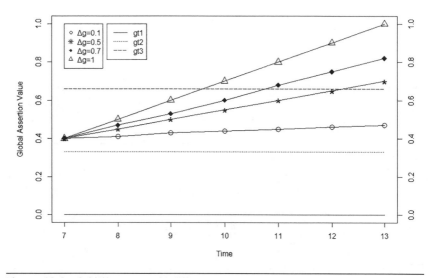

Figure 10.2 S-SAKA response with different Δg settings.

Although the computation and execution times of mobile sink, cluster head and base stations are lower than the proposed protocol, total execution time in term of milliseconds the proposed S-SAKA performs lowest time. It is also remarked that S-SAKA can divide the protected resources in IoT-based environments into a number of assertion levels ($= n$). Assuming a total count of assertion levels n equal to 3, threshold values of the assertion levels can be as shown in Table 10.4.

$g_{ti} \in [0,1]$ and represents the assertion level threshold value of the ith assertion level that can be calculated by $g_{ti} = (i-1) * \dfrac{1}{n}$. g_{ti} reflects how confident the S-SAKA system must be about a user in order to assert his/her identity before granting access to them. It defines a control parameter representing the rate of change of the assertion value and referred to as Δg. This parameter can be used to control the speed by which the assertion value in safety-inspired applications' increases or decreases.

Changing the value of Δg affects how the system confidence is about a user access as shown in Figure 10.2. In this figure when Δg is set to values between 0.1 and 0.5, the S-SAKA system confidence increases slowly and needs at least 6 events for the global assertion value to reach the next threshold value. This setting would be useful in IoT environments, where high security levels are a must such as the case in safety-inspired applications. However, when Δg has high values such 0.7 to 1.0, the system confidence rises much faster with less number of events to reach the second level; this setting would be useful for more relaxed IoT environments.

VII. CONCLUSION

In this paper, WSN security schemes are considered in terms of authentication and secure key agreement, which can be essential particularly for the IoT applications for public safety applications with cloud interactions. Enhancement of security framework can be essential for public safety paradigm since the IoT systems can be used for communication of sensitive information. Addressing the potential security based challenges, Seamless secure authentication and key agreement (S-SAKA) framework using bilinear-pairing and elliptic-curve cryptosystems has been proposed for the security issues, like data confidentiality, mutual authentication, session-key agreement, user anonymity, intractability and resilient to node-capture, key impersonation, password guessing and stolen smart-card attack. While using mobile-sink in WSNs, the S-SAKA framework does not only solve some major security issues, but also ensures a seamless connectivity to reduce the computation and communication cost of the network systems. In terms of authentication and authorization, recent studies on formal verifications that are based on bilinear pairing and elliptic-curve cryptosystems [44] are not provided for WSNs. As stated earlier, considering life-time of WSN, security aspects are critical and new solutions should be provided effectively and efficiently. The formal verification method and critical analyses performed prove that proposed S-SAKA provides mutual authentication, secure key agreement and data confidentiality. Furthermore, the results of performance evaluation show the reduced overhead of the proposed approach compared to the existing studies. Thus, the proposed S-SAKA framework can be well suited to the environments where public safety networks make use of IoT based applications with wireless sensors networks.

REFERENCES

[1] L. Atzori, A. Iera, and G. Morabito, "The internet of things: a survey," *Comput Netw*, vol. 54, no. 15, pp. 2787–805, Oct. 2010.

[2] O. Vermesan, and P. Friess, Internet of things-global technological and societal trends from smart environments and spaces to green ICT. River Publishers, Denmark, 2011.

[3] D. Minoli, S. Kazem, and B. Occhiogrosso. "IoT Considerations, Requirements, and Architectures for Smart Buildings—Energy Optimization and Next-Generation Building Management Systems," IEEE Internet Things J, vol. 4, no. 1, pp. 269-283, Feb. 2017.

[4] I. Butun, M. Erol-Kantarci, B. Kantarci, and H. Song, "Cloud-centric multi-level authentication as a service for secure public safety device networks," IEEE Commun. Mag., vol. 54, no. 4, pp. 47–53, Apr.2016.

[5] M. Kamruzzaman, N. I. Sarkar, J. Gutierrez, and S. K. Ray, "A study of IoT-based post-disaster management," in Proc. ICOIN, Da Nang, Vietnam, 2017, pp. 406–410.

[6] Z. Chu, H. X. Nguyen, T. Anh Le, M. Karamanoglu, D. To, E. Ever, F. Al-Turjman, and Adnan Yazici, "Game theory based secure wireless powered D2D communications with cooperative jamming," in Wireless Days, Porto, Portugal, 2017, pp. 95-98.

[7] G. Solmaz, and D. Turgut, "Event coverage in theme parks using wireless sensor networks with mobile sinks," in Proc. ICC, Budapest, Hungary, 2013, pp. 1522–1526.

[8] R. Rahmatizadeh, S. Khan, A.P. Jayasumana, D. Turgut, and L. Bölöni, "Routing towards a mobile sink using virtual coordinates in a wireless sensor network," in Proc. ICC, Sydney, NSW, Australia, 2014, pp. 12–17.

[9] J. Yick, B. Mukherjee, and D. Ghoshal, "Wireless sensor network survey," Comput. Netw, vol. 52, no.12, pp. 2292–2330, Aug. 2008.

[10] S. Kumari, M. K. Khan, and M. Atiquzzaman, "User authentication schemes for wireless sensor networks: a review," Ad Hoc Netw, vol. 27, no: C, pp. 159–194, Apr. 2015.

[11] D. He, N. Kumar, and N. Chilamkurti, "A secure temporal-credential-based mutual authentication and key agreement scheme with pseudo identity for wireless sensor networks," Info. Sci., vol. 321, pp. 263–277, Nov. 2015.

[12] D. He, and D. Wang, "Robust biometrics-based authentication scheme for multi- server environment," IEEE Syst. J. vol. 9, no. 3, pp. 816–823, Sept. 2015.

[13] D. Wang, and P. Wang, "Understanding security failures of two-factor authentication schemes for real-time applications in hierarchical wireless sensor networks," Ad Hoc Netw., vol. 20, pp. 1–15, Sept.2014.

[14] D. Wang, and P. Wang, "On the usability of two-factor authentication," in Proc. Secure Comm., Beijing, China, 2014, pp. 141–150.

[15] D. Wang, D. He, P. Wang, and C.H. Chu, "Anonymous two-factor authentication in distributed systems: certain goals are beyond attainment," IEEE Trans Dependable Secure Comput., vol. 12, no. 4, pp. 428–442, Jul-Aug. 2015.

[16] K. H. Wong, Y. Zheng, J. Cao, and S. Wang, "A dynamic user authentication scheme for wireless sensor networks," in Proc. SUTC'06, Taiwan, 2006, pp. 244–251.

[17] B. D. Deebak, "Secure and efficient mutual adaptive user authentication scheme for heterogeneous wireless sensor networks using multimedia client–server systems," Wireless Pers. Commun., vol. 87, no. 3, pp. 1013–1035, Apr. 2016.

[18] M. Turkanovi'c, B. Brumen, and M. Hölbl, "A novel user authentication and key agreement scheme for heterogeneous ad hoc wireless sensor networks, based on the internet of things notion," Ad Hoc Netw., vol. 20, pp. 96–112, Sept. 2014.

[19] M. S. Farash, M. Turkanovi'c, S. Kumari, and M. Hölbl, "An efficient user authentication and key agreement scheme for heterogeneous wireless sensor network tailored for the internet of things environment," Ad Hoc Netw., vol. 36, pp. 152–176, Jan. 2016.

[20] A.K. Das, A.K. Sutrala, S. Kumari, V. Odelu, M. Wazid, and X. Li, "An efficient multi–gateway-based three-factor user authentication and key agreement scheme in hierarchical wireless sensor networks," Secur. Commun. Netw., vol. 9, no. 13, pp. 2070–2092, Sept. 2016.

[21] R. Amin, and G. Biswas, "A secure light weight scheme for user authentication and key agreement in multi-gateway based wireless sensor networks," Ad Hoc Netw., vol. 36, pp. 58–80, Jan. 2016.

[22] J. Srinivas, S. Mukhopadhyay, and D. Mishra, "Secure and efficient user authentication scheme for multi-gateway wireless sensor networks," Ad Hoc Netw., vol. 54, pp. 147–169, Jan. 2017.

[23] A.K. Das, P. Sharma, S. Chatterjee, and J.K. Sing, "A dynamic password-based user authentication scheme for hierarchical wireless networks," J Netw. Comput. Appl. vol. 35, no. 5, pp. 1646–1656, Sept. 2012.

[24] D. He, "An efficient remote user authentication and key agreement protocol for mobile client-server environment from pairings," Ad Hoc Netw., vol. 10, no. 6, pp. 1009–1016, Aug. 2012.

[25] M. Burrows, M. Abadi, and R. Needham, "A logic of authentication," ACM Trans. Comput. Syst, vol. 8, no. 1, pp. 18–36, Feb. 1990.

[26] D. Wang, N. Wang, P. Wang, and S. Qing, "Preserving privacy for free: efficient and provably secure two-factor authentication scheme with user anonymity," Info. Sci. vol. 321, pp. 162–178, Nov. 2015.

[27] M. L. Das, "Two-factor user authentication in wireless sensor networks," IEEE Trans. Wireless Commun., vol. 8, no.3, pp. 1086–1090, Mar. 2009.

[28] J. Yuan, C. Jiang, and Z. Jiang, "A biometric-based user authentication for wireless sensor networks", Wuhan University Journal of Natural Sciences, vol. 15, no. 3, pp. 272–276, Jun. 2010.

[29] H. R. Tseng, R. H. Jan, and W. Yangand, "An improved dynamic user authentication scheme for wireless sensor networks," in Proc. GLOBECOM, Washington, DC, USA, 2007, pp. 986–990.

[30] T. H. Lee, "Simple dynamic user authentication protocols for wireless sensor networks," in Proc. SENSORCOMM, Cap Esterel, France, pp. 657–660, 2008.

[31] E. G. AbdAllah, H. S. Hassanein, and M. Zulkernine, "A Survey of Security Attacks in Information-Centric Networking," IEEE Commun. Surveys Tuts., vol. 7, no. 3, pp. 1441–1454, Jan. 2015.

[32] L. Eschenauer, and V. D. Gligor, "A key-management scheme for distributed sensor networks," in Proc. CCS, Washington, DC, USA, 2002, pp. 41–47.

[33] H. Chan, A. Perrig, and D. Song, "Random key pre-distribution schemes for sensor networks," in Proc. SSP, Oakland, California, USA, 2003, pp. 197–213.

[34] A. Rasheed, and R. N. Mahapatra, "The three-tier security scheme in wireless networks with mobile sinks," IEEE Trans. Parallel Distrib. Syst., vol. 23, no. 5, pp. 958–965, May 2012.

[35] R. Watro, D. Kong, S. F. Cuti, C. Gardiner, C. Lynn, and P. Kruus, "TinyPK: securing sensor networks with public key technology," in Proc. SASN, Washington, DC, USA, 2004, pp. 59–64.

[36] T. H. Chen, and W. K. Shih, "A robust mutual authentication protocol for wireless sensor networks," ETRI Journal, vol. 32, no.5, pp. 704–712, Oct. 2010.

[37] D. He, Y. Gao, S. Chan, C. Chen, and J. Bu, "An enhanced two factor user authentication scheme in wireless sensor networks," Ad-Hoc & Sensor Wireless Netw., vol. 10, no. 4, pp. 361–371, Feb. 2010.

[38] S. Park, B. Aslam, D. Turgut, and C. Zou, "Defense against Sybil attack in the initial deployment stage of vehicular ad hoc network based on roadside unit support," Secur. Commun. Netw., vol. 6, no. 4, pp. 523–538, Jan. 2013.

[39] O. Delgado-Mohatar, A. Fuster-Sabater, and J. M. Sierra, "A lightweight authentication scheme for wireless sensor networks," Ad Hoc Netw., vol. 9, no. 5, pp. 727–735, Jul. 2011.

[40] A. D. Karaoğlan, A. B. Muhammed, L. Albert, and S. Erkay, "DKEM: Secure and efficient Distributed Key Establishment Protocol for Wireless Mesh Networks," Ad Hoc Netw., vol. 54, pp. 53–68, Jan. 2017.

[41] W. Shi, and P. Gong, "A new user authentication protocol for wireless sensor networks using elliptic curves cryptography," Distrib. Sensor Netw., pp.1-7, Jan. 2013.

[42] T. Bonaci, L. Bushnell, and R. Poovendran, "Node capture attacks in wireless sensor networks: A system theoretic approach," in Proc. CDC, Atlanta, Georgia, USA, 2010, pp. 6765–6772.

[43] T. M. Vu, R. Safavi-Naini, and C. Williamson, "Securing wireless sensor networks against large-scale node capture attacks," in Proc. ASIACCS, Beijing, China, 2010, pp. 112–123.

[44] D. Hankerson, A. Menezes, and S. Vanstone, Guide to Elliptic Curve Cryptography, New York, NY, USA: Springer, 2004.

[45] L. C. Washington, Elliptic Curves Number Theory and Cryptography, 2nd ed., Boca Raton, FL, USA: Chapman and Hall/CRC, 2008.

[46] N. Kumar, S. Misra, N. K. Chilamkurti, J. Lee, and J. J. P. C. Rodrigues, "Bayesian Coalition Negotiation Game as a Utility for Secure Energy Management in a Vehicles-to-Grid Environment," IEEE Trans. Dependable Sec. Comput., vol. 13, no. 1, pp. 133–145, Jan-Feb. 2016.

[47] F. Al-Turjman, M. Imran, and A. Vasilakos, "Value-based Caching in Information-Centric Wireless Body Area Networks," Sensors J., vol. 17, no. 1, pp. 1-19, Jan. 2017.

[48] C. Stergiou, K. E. Psannis, B. G. Kim, and B. Gupta, "Secure integration of IoT and Cloud Computing," Future Gener Comput Syst to be published. DOI: 10.1016/j.future.2016.11.031.

[49] M. Tao, J. Zuo, Z. Liu, A. Castiglione, and F. Palmieri, "Multi-layer cloud architectural model and ontology-based security service framework for IoT-based smart homes," Future Gener Comput Syst, to be published. DOI: 10.1016/j.future.2016.11.011.

[50] B. Kantarci, and H. T, Mouftah, "Trustworthy sensing for public safety in cloud-centric internet of things," IEEE Internet Things J., vol. 1, no.4, pp. 360–368, Aug. 2014.

Index

SECOND EDITION

Food Plant
SANITATION
Design, Maintenance, and Good Manufacturing Practices